GEOTECHNICAL SPECIAL PUBLICATION NO. 76

GEOSYNTHETICS IN FOUNDATION REINFORCEMENT AND EROSION CONTROL SYSTEMS

PROCEEDINGS OF SESSIONS OF GEO-CONGRESS 98

SPONSORED BY
The Geosynthetics Committee of
The Geo-Institute of
The American Society of Civil Engineers

October 18–21, 1998
Boston, Massachusetts

EDITED BY
John J. Bowders
Heather B. Scranton
Gregory P. Broderick

WITHDRAWN

1801 ALEXANDER BELL DRIVE
RESTON, VIRGINIA 20191–4400

Abstract: The variety and number of applications of geosynthetics in geotechnical engineering continues to grow rapidly. This proceedings focuses on two specific applications: soil reinforcement beneath shallow foundations and erosion control systems. The contributions are from academicians, practicing engineers, government engineers, and manufacturers. Topics addressed in the foundation reinforcement papers include: 1) analytical procedures to determine the bearing capacity of a geosynthetic-reinforced foundation; 2) laboratory investigations of geogrid reinforced soils examining stress distribution under static loads and settlement as a result of cyclic loading; 3) full-scale field studies of preloaded/prestressed geosynthetic-reinforced soils; and 4) bearing capacity and settlement design practices of spread footings over geosynthetic-reinforced foundations. Topics addressed in the geosynthetic erosion control papers include a field study of a landfill cover, a case study of a biotechnically stabilized earthen buttress, and a case study of a vegetative channel lining. An erosion control materials test facility and vegetation selection for geosynthetic erosion control products are addressed in two additional papers.

Library of Congress Cataloging-in-Publication Data

Geo-Congress (1998: Boston, Mass.)
    Geosynthetics in foundation reinforcement and erosion control systems: proceedings of the sessions of the Geo-Congress sponsored by the Geosynthetics Committee of the Geo-Institute of the American Society of Civil Engineers: Boston, Massachusetts, October 18-21, 1998 / edited by John J. Bowders, Heather B. Scranton, and Gregory P. Broderick.
    p.   cm.   –(Geotechnical special publication)
    Includes bibliographical references and index.
    ISBN 0-7844-0383-X
    1. Foundations–Congresses. 2. Geosynthetics–Congresses. 3. Soil stabilization–Congresses. 4. Soil conservation–Congresses. I. Bowders, John J. II. Scranton, Heather B. III. Broderick, Gregory P. IV. American Society of Civil Engineers. Geo-Institute. Geosynthetics Committee. V. Title. VI. Series.
    TA775.G35   1998   98-38065
    624.1'53–dc21   CIP

Any statements expressed in these materials are those of the individual authors and do not necessarily represent the views of ASCE, which takes no responsibility for any statement made herein. No reference made in this publication to any specific method, product, process or service constitutes or implies an endorsement, recommendation, or warranty thereof by ASCE. The materials are for general information only and do not represent a standard of ASCE, nor are they intended as a reference in purchase specifications, contracts, regulations, statutes, or any other legal document.

ASCE makes no representation or warranty of any kind, whether express or implied, concerning the accuracy, completeness, suitability, or utility of any information, apparatus, product, or process discussed in this publication, and assumes no liability therefore. This information should not be used without first securing competent advice with respect to its suitability for any general or specific application. Anyone utilizing this information assumes all liability arising from such use, including but not limited to infringement of any patent or patents.

Photocopies. Authorization to photocopy material for internal or personal use under circumstances not falling within the fair use provisions of the Copyright Act is granted by ASCE to libraries and other users registered with the Copyright Clearance Center (CCC) Transactional Reporting Service, provided that the base fee of $8.00 per chapter plus $.50 per page is paid directly to CCC, 222 Rosewood Drive, Danvers, MA 01923. The identification for ASCE Books is 0-7844-0383-X/98/ $8.00 + $.50 per page. Requests for special permission or bulk copying should be addressed to Permissions & Copyright Dept., ASCE.

Copyright © 1998 by the American Society of Civil Engineers,
All Rights Reserved.
Library of Congress Catalog Card No: 98-38065
ISBN 0-7844-0383-X
Manufactured in the United States of America.

Geotechnical Special Publications

1. *Terzaghi Lectures*
2. *Geotechnical Aspects of Stiff and Hard Clays*
3. *Landslide Dams: Processes, Risk, and Mitigation*
4. *Tiebacks for Bulkheads*
5. *Settlement of Shallow Foundation on Cohesionless Soils: Design and Performance*
6. *Use of In Situ Tests in Geotechnical Engineering*
7. *Timber Bulkheads*
8. *Foundations for Transmission Line Towers*
9. *Foundations & Excavations in Decomposed Rock of the Piedmont Province*
10. *Engineering Aspects of Soil Erosion, Dispersive Clays and Loess*
11. *Dynamic Response of Pile Foundations –Experiment, Analysis and Observation*
12. *Soil Improvement: A Ten Year Update*
13. *Geotechnical Practice for Solid Waste Disposal '87*
14. *Geotechnical Aspects of Karst Terrains*
15. *Measured Performance Shallow Foundations*
16. *Special Topics in Foundations*
17. *Soil Properties Evaluation from Centrifugal Models*
18. *Geosynthetics for Soil Improvement*
19. *Mine Induced Subsidence: Effects on Engineered Structures*
20. *Earthquake Engineering & Soil Dynamics II*
21. *Hydraulic Fill Structures*
22. *Foundation Engineering*
23. *Predicted and Observed Axial Behavior of Piles*
24. *Resilient Moduli of Soils: Laboratory Conditions*
25. *Design and Performance of Earth Retaining Structures*
26. *Waste Containment Systems: Construction, Regulation, and Performance*
27. *Geotechnical Engineering Congress*
28. *Detection of and Construction at the Soil/Rock Interface*
29. *Recent Advances in Instrumentation, Data Acquisition and Testing in Soil Dynamics*
30. *Grouting, Soil Improvement and Geosynthetics*
31. *Stability and Performance of Slopes and Embankments II*
32. *Embankment of Dams –James L. Sherard Contributions*
33. *Excavation and Support for the Urban Infrastructure*
34. *Piles Under Dynamic Loads*
35. *Geotechnical Practice in Dam Rehabilitation*
36. *Fly Ash for Soil Improvement*
37. *Advances in Site Characterization: Data Acquisition, Data Management and Data Interpretation*
38. *Design and Performance of Deep Foundations: Piles and Piers in Soil and Soft Rock*

39	*Unsaturated Soils*
40	*Vertical and Horizontal Deformations of Foundations and Embankments*
41	*Predicted and Measured Behavior of Five Spread Footings on Sand*
42	*Serviceability of Earth Retaining Structures*
43	*Fracture Mechanics Applied to Geotechnical Engineering*
44	*Ground Failures Under Seismic Conditions*
45	*In Situ Deep Soil Improvement*
46	*Geoenvironment 2000*
47	*Geo-Environmental Issues Facing the Americas*
48	*Soil Suction Applications in Geotechnical Engineering*
49	*Soil Improvement for Earthquake Hazard Mitigation*
50	*Foundation Upgrading and Repair for Infrastructure Improvement*
51	*Performance of Deep Foundations Under Seismic Loading*
52	*Landslides Under Static and Dynamic Conditions–Analysis, Monitoring, and Mitigation*
53	*Landfill Closures–Environmental Protection and Land Recovery*
54	*Earthquake Design and Performance of Solid Waste Landfills*
55	*Earthquake-Induced Movements and Seismic Remediation of Existing Foundations and Abutments*
56	*Static and Dynamic Properties of Gravelly Soils*
57	*Verification of Geotechnical Grouting*
58	*Uncertainty in the Geologic Environment*
59	*Engineered Contaminated Soils and Interaction of Soil Geomembranes*
60	*Analysis and Design of Retaining Structures Against Earthquakes*
61	*Measuring and Modeling Time Dependent Soil Behavior*
62	*Case Histories of Geophysics Applied to Civil Engineering and Public Policy*
63	*Design with Residual Materials: Geotechnical and Construction Considerations*
64	*Observation and Modeling in Numerical Analysis and Model Tests in Dynamic Soil-Structure Interaction Problems*
65	*Dredging and Management of Dredged Material*
66	*Grouting: Compaction, Remediation and Testing*
67	*Spatial Analysis in Soil Dynamics and Earthquake Engineering*
68	*Unsaturated Soil Engineering Practice*
69	*Ground Improvement, Ground Reinforcement, Ground Treatment: Developments 1987-1997*
70	*Seismic Analysis and Design for Soil-Pile-Structure Interactions*
71	*In Situ Remediation of the Geoenvironment*
72	*Degradation of Natural Building Stone*
73	*Innovative Design and Construction for Foundations and Substructures Subject to Freezing and Frost*
74	*Guidelines of Engineering Practice for Braced and Tied-Back Excavations*
75	*Geotechnical Earthquake Engineering and Soil Dynamics III*
76	*Geosynthetics in Foundation Reinforcement and Erosion Control Systems*

PREFACE

Heretofore, geosynthetics were treated as a subset of the Soil Improvement Committee. Noting the wide variety of applications of geosynthetics, the Geo-Institute decided to establish a separate committee in this area of specialization. These proceedings and the associated sessions at the Geo-Congress mark the initial efforts of the recently formed Geosynthetics Committee of the Geo-Institute.

The sessions were conceived by the members of the G-I Geosynthetics committee. More specifically, Gregory P. Broderick, David J. Elton, Mo A. Gabr, and Heather B. Scranton developed the session topics, solicited the papers, managed the review process and chaired the sessions. John Bowders served as the overall coordinator of the process and co-editor of this volume which marks the first publication by the G-I Geosynthetics committee.

These proceedings focus on geosynthetic soil reinforcement beneath shallow foundations and geosynthetic erosion control systems. The contributions are from academicians, practicing engineers, government engineers and manufacturers. Topics addressed in the foundation reinforcement papers include: analytical procedures to determine the bearing capacity of a geosynthetic-reinforced foundation, laboratory investigations of geogrid reinforced soils examining stress distribution under static loads and settlement as a result of cyclic loading, full-scale field studies of preloaded/prestressed geosynthetic-reinforced soils, and bearing capacity and settlement design practices of spread footings over geosynthetic-reinforced foundations. Topics addressed in the geosynthetic erosion control papers include: a field study of a landfill cover, a case study of a biotechnically stabilized earthen buttress, and a case study of a vegetative channel lining. An erosion control materials test facility and vegetation selection for geosynthetic erosion control products are addressed in two additional papers.

All of the papers presented in the geosynthetics sessions at Geo-Congress and published in this volume have been peer-reviewed consistent with the standards of the ASCE's *Journal of Geotechnical and Geoenvironmental Engineering*. Each paper was revised to conform to mandatory revisions of the reviewers and received positive peer reviews before ultimately being published. All papers are eligible for discussion in The Journal and are eligible for ASCE awards.

Paper reviews were anonymous and the quality of the proceedings reflects the efforts of the reviewers listed below. The editors extend their sincerest appreciation to the reviewers for their selfless, accurate and timely efforts towards this proceeding.

July 1998

John J. Bowders
University of Missouri – Columbia
Columbia, Missouri

Heather B. Scranton
Haley and Aldrich, Inc.
Boston, Massachusetts

Gregory P. Broderick
University of New Haven
New Haven, Connecticut

Paper Reviewers for *Geosynthetics in Foundation Reinforcement and Erosion Control Systems*. * GI-Geosynthetics Committee member, [+]Multiple papers

Ryan R. Berg
James G. Collin*
Albert F. DiMillio
Richard Goodrum
Ganesh Gopalakrishnan
Robert D. Holtz
Michael Houlihan
Hoe I. Ling*
Mark S. Meyers*
Radoslaw L. Michalowski
Bob Moran
Gerald P. Raymond
Richard A. Reid*
Cetin Soydemir
C. Joel Sprague*[+]
Marc Theisen
Jonathan T.H. Wu

Contents

Geosynthetics in Foundation Reinforcement

The Design of Geosynthetic Reinforced Foundations 1
    Mark H. Wayne, Jie Han, and Ken Akins

Dynamic Loading on Foundation on Reinforced Sand 19
    B. M. Das

Behavior of the First Prototype and Full-Scale Models of PLPS
Geosynthetic-Reinforced Soil Structure 34
    Taro Uchimura, Fumio Tatsuoka, Masaru Tateyama, and Tetsushi Koga

Slip-Line Analyses of Geosynthetic-Reinforced Strip Footings 49
    Aigen Zhao

A Study of Stress Distribution in Geogrid-Reinforced Sand 62
    M. A. Gabor, Robert Dodson, and James G. Collin

Geosynthetic Erosion Control Systems

Geosynthetic Erosion Control Materials: A Landfill Cover Field Study 77
    George R. Koerner and David A. Carson

Partnering Geosynthetics and Vegetation for Erosion Control 92
    Robbin B. Sotir, John T. Difini, and Andrew F. McKown

Performance Testing of Erosion Control BMPs at the ErosionLab 103
    Dwight Cabalka and Paul Clopper

Geosynthetically Reinforced Vegetation: Providing an Effective, Economical and
Aesthetically Pleasing Alternative to Rock Riprap 116
    Roy J. Nelsen

Vegetation Selection for Rolled Erosion Control Product 130
    Daniel Hunt, Deron N. Austin, and William Agnew

Subject Index 145
Author Index 147

The Design of Geosynthetic Reinforced Foundations

Mark H. Wayne[1], Jie Han[2], and Ken Akins[3]

Abstract

In an effort to improve the performance of poor soils geotechnical engineers are faced with four potential solutions. These include: (1) replace the poor ground with a stronger material or combination of materials, (2) alter the natural condition of the poor soil to meet project requirements, (3) redesign the structure to meet the soil limitations, or (4) bypass the unsuitable soil laterally through relocation of the facility, or vertically through the use of a deep foundation. The purpose of this paper is to examine the first option. In particular, this paper reviews the work that has been conducted to date on this topic as it relates to the inclusion of geosynthetics as reinforcement within the replaced zone. This information is then used to examine the influence that punched and drawn biaxial geogrids have on the design of a geosynthetic reinforced foundation (GRF) with an emphasis on bearing capacity and settlement of spread footings. The paper provides typical design parameters for these geogrids used in reinforced foundations.

Introduction

Throughout the world developers have discovered that some of the most ideal locations for their clients are situated within land lots containing poor near surface soils. These developers are often faced with the traditional solutions that either involve redesign of the building to meet the soil limitations or the use of a deep foundation system. Because the costs associated with redesign are often prohibitive, designers often choose the latter option. Use of the deep foundation

---
[1] Member, Manager of Technology Development, Tensar Earth Technologies, Inc., 5775-B Glenridge Dr., Suite 450, Atlanta, GA 30328
[2] Associate Member, Senior Engineer, Tensar Earth Technologies, Inc., 5775-B Glenridge Dr., Suite 450, Atlanta, GA 30328
[3] Member, Manager of Engineering, Tensar Earth Technologies, Inc., 5775-B Glenridge Dr., Suite 450, Atlanta, GA 30328

approach allows designers to transfer a large portion of building loads to better bearing materials below the poor near surface soils. However, these systems are expensive, and even with the development and use of more economical types of deep foundation systems, developers are demanding more economical solutions to this problem. Numerous ground modification techniques have been developed within the past quarter of a century to meet this growing demand. The shared commonality of these solutions is related to modification of the engineering properties of the near surface soils such that economical shallow foundation systems can be utilized in construction of the facility. Munfakh (1996) has divided ground modification techniques into categories which include: (1) Densification, (2) Consolidation, (3) Reinforcement, (4) Chemical Stabilization, (5) Thermal Stabilization, and (6) Biotechnical Stabilization. In order to properly improve the condition of near surface soils, these techniques must be able to: 1.) increase the bearing capacity; 2.) control deformations, or 3.) increase resistance to liquefaction.

This paper explores the design of geosynthetic reinforced foundations as a means of ground modification. A complete overview of research work is presented to give the reader a better understanding of the benefits that can be derived through the use of this ground modification technique. Emphasis is then placed on the examination of those factors which control the design of a geosynthetic reinforced foundation. Finally, design guidance is suggested by providing typical design parameters for geogrids used in reinforced foundations.

Literature Review

There is a great deal of literature which focuses on the use of geosynthetics as a plausible means of increasing the bearing capacity of soil. The improvement in bearing capacity is commonly expressed as the ratio of the ultimate bearing pressure for the reinforced soil to that of the unreinforced soil, and is referred to as the ultimate bearing capacity ratio or BCR. Some researchers have even examined an allowable BCR which defines the improvement in bearing capacity at a predefined level of settlement. All the BCR values cited in this paper are referred as the ultimate bearing capacity ratio unless they are specified. A large number of papers evaluated the results of testing on model footings. Based on the scale effect tests conducted by Das (1994) using different sizes of model strip footings on unreinforced and reinforced sands, he concluded that "...the magnitude of B (footing width) in the laboratory model tests should be equal to or greater than 120 mm." It was observed that the bearing capacity ratio (BCR) reached a constant value for model footing widths greater than or equal to 120 mm.

Research work conducted by Guido, et al. (1985, 1986, 1987), and Fragaszy and Lawton (1985) clearly demonstrates the fact that both the type of reinforcement, (metal, geotextile, or geogrid) and manner in which the reinforcement is incorporated within the soil is directly related to the engineering properties achieved by the combined materials. More important, these investigators discovered different ranges of improvement for each type of reinforcement. In particular, Guido (1986) reported that for geotextiles the BCR varied from 1.65 to 1.75, and for punched and drawn biaxial geogrids this ratio varied from 2 - 3. More importantly, the minimum ratio of reinforcement width to footing width (b/B ratio) required to get a significant benefit from the GRF was found to be 3.25 for geotextiles versus 1.5 for the punched and drawn biaxial geogrids used in his study. Fragaszy and Lawton (1985) examined the behavior of aluminum foil as a reinforcing mechanism. They reported that for metal reinforcement the ultimate bearing capacity ratio varied from 1.2 - 1.7. For all of the reinforcements evaluated the variation in BCR was a function of geometric and mechanical properties of the reinforcement and engineering properties of the soil used to construct the reinforced soil mass. As such, selection of the proper reinforcing material and configuration of this material within the soil are both influential in achieving the desired end effect.

Gabr, et al. (1996) examined both bearing capacity and settlement for two types of punched and drawn biaxial geogrid, a well graded sand, and a sandy gravel. Plate loading was performed on 1.2 m deep of soil in a 1.5 m square box using a 0.3 m square metal plate. These researchers observed a stiffer response for s/B (settlement/footing width)= .03 and .015 strain deformation level as compared to the unreinforced case. An allowable BCR range of 1.25 - 1.6 was achieved at a deformation of 9 mm and an allowable BCR range of 1.23 - 2.2 was achieved at a deformation of 4.5 mm. The author's also noticed a consistent improvement in the level and consistency of the dry unit weight with depth. Further testing by Gabr (1997) was conducted using strain gages and total earth pressure transducers within the reinforced zone. Strain readings in the geogrid under a load in excess of 430 kPa reached 0.5% for a single layer system, 0.35% for a two layer system, and 0.25% for a three layer system.

In addition to laboratory studies, there were several researchers who conducted relatively large scale field studies. The first large scale study was conducted by Chadbourne (1994) who reported the results of large scale plate load tests, 0.3 m by 0.3 m, 0.45 m by 0.45 m, 0.6 m by 0.6 m, and 0.9 m by 0.9 m square footings, on sand with and without geosynthetic reinforcement. Both the punched and drawn geogrid and a geocell were examined in this test program. The parameters varied in the reinforcement portion of the test program included: 1) number of reinforcement layers (N), 2) the width dimension of the square reinforcing layer (b/B ratio), and 3) the distance between the uppermost layer of reinforcement and the bottom of the footing (u/B ratio). This researcher examined the response of one and two layers of geogrid and one geocell layer. He then compared these responses against the unreinforced case. This researcher found that the ultimate bearing capacity ratio varied from 2.7 to 3.25 for a 0.9 m by 0.9 m, and 0.3 m by 0.3 m spread footing respectively.

Adams and Collin (1997) provide a comprehensive review of data from thirty four large scale model load tests. A series of these large scale plate load tests were conducted with one, two, and three layers of a punched and drawn biaxial geogrid and one geocell. Tests included 0.3 m by 0.3 m, 0.45 m by 0.45 m, 0.6 m by 0.6 m, and 0.9 m by 0.9 m square footings, on a poorly graded concrete mortar sand with and without geosynthetic reinforcement. The parameters varied in the reinforcement portion of the test program included: 1) number of reinforcement layers (N), 2) the plan dimensions of the square reinforcing layer (b/B ratio), and 3) the density of the soil. These researchers examined the response of 3 layers of geogrid and one layer of geocell and compared these responses against the unreinforced case. These researchers verified that 1.) for the biaxial geogrid an average increase of 230% (i.e., $1.92 \leq BCR \leq 2.63$) in allowable bearing capacity was achieved at settlements of 10 mm and 20 mm, respectively. An allowable bearing capacity ratio of 2.63 was achieved for the GRF versus an ultimate bearing capacity ratio of 1.27 for the geocell. The researchers pointed out the fact that a considerable amount of deformation was required before any improvement could be realized from the geocell.

Tsukada, et. al. (1993) explored the use of punched and drawn biaxial geogrids for construction of a roadway foundation. They examined the response of two different GRF systems when placed over soil-cement columns. The GRF's were composed of either 1.) single and double layers of geogrid or 2.) two different thickness' of a geogrid GRF. They found that the settlement response and pressure distribution was directly related to the thickness and configuration of the GRF. In fact, the vertical pressure at the top of the soil-cement columns was reduced by using two geogrid layers to 2/3rd of the value obtained with a single layer.

Ochiai, et. al., (1993) used a laboratory setup consisting of a series of springs with an overlying geogrid GRF. A 0.1 m wide by 0.4 m long rectangular plate was used to apply the vertical loads. The subgrade stiffness was varied from 1098 kPa/m to 5018 kPa/m. The concept behind this work was directly related to the evaluation of the stress distribution under a geogrid GRF overlying subgrades exhibiting varying degrees of stiffness. These author's found "that the width over which the vertical stress is distributed increases as the thickness of the geogrid GRF becomes greater, and the supporting foundation with lower vertical stiffness (lower modulus) tends to offer a wider vertical stress distribution...". This work also verified the field behavior of the single, multilayer, and mattress system constructed by Tsukada, et. al.(1993).

In summary, the influence that a geosynthetic inclusion has on bearing capacity and settlement must be understood prior to design of a geosynthetic reinforced foundation. As such, both of these issues are discussed in the sections which follow.

Bearing Capacity Design Issues

Calculation of bearing capacity of a foundation is generally regarded as a first step for foundation design in practice. When the design load from a superstructure is less than the allowable bearing capacity, an engineer certain about the foundation will have a certain factor of safety in service. For this case, the load - deformation behavior is nearly linear. Hence, elastic theory can be reasonably adopted to estimate the settlement of a foundation. For a native foundation, several approaches are available to estimate the ultimate bearing capacity of a foundation. The most commonly used methods were proposed by Terzaghi (1943), Meyerhof (1951, 1963), and Hansen (1957, 1970). For a reinforced foundation, many model tests have been performed to study influence factors on bearing capacity including the distance of the uppermost reinforcement to the bottom of the footing (u), the spacing between reinforcements (s), the width of reinforcements (b), the number of reinforcements (N), the thickness of the reinforced soil (z), and the type of reinforcements. At least four possible failure modes exist for reinforced foundations as shown in Fig. 1. Different configurations of reinforcements and soil conditions may result in different possible failure modes. The goal in design is to determine the controlling failure mode.

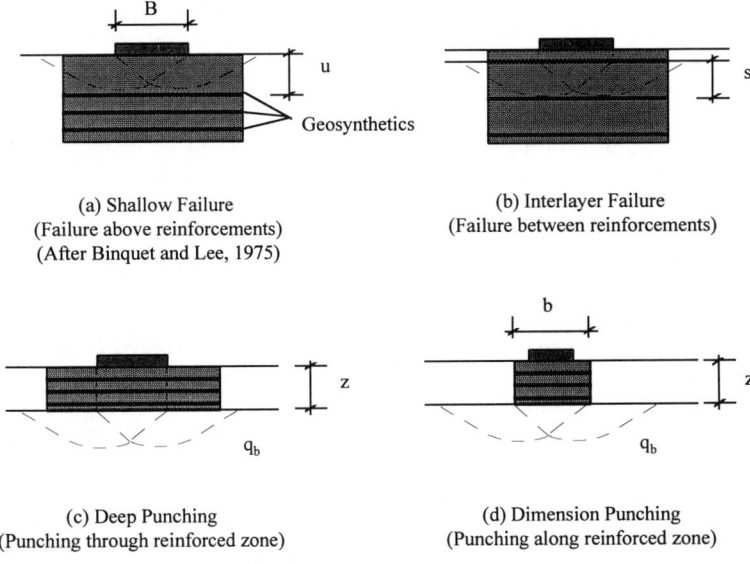

Figure 1. Possible Failure Mechanisms for the Geosynthetic Reinforced Foundation (GRF)

Failure above the uppermost reinforcement

The failure above the uppermost reinforcement, as depicted in Figure 1 (a) can be easily prevented by placing the uppermost reinforcement close enough to the footing. The maximum distance from the bottom of the footing to the uppermost reinforcement can be approximately estimated by using the approach proposed by Mandel and Salencon (1972) and Bonaparte and Christopher (1987). Although these approaches was originally developed for calculating the bearing capacity of the foundation or embankment with an existence of a rigid base, the comparison in Fig. 2(a) demonstrated that the test data for reinforced sands approximately fall within the trend of the theoretical results when the ratio of the distance to the uppermost reinforcement to the width of the footing (u/B) increases. It is observed that measured BCR values are less than calculated values when u/B is relatively small. The reason is that other failure mechanism occurs prior to this failure mechanism. Based on limited information, however, the test data for reinforced clays, as depicted in Fig. 2(b), have shown measured BCR values greater than calculated values when u/B is relatively small. The reason for this result is unclear. In Fig. 2(b), rough or smooth interface is referred as the interface behavior between soft soil and stiff soil. Binquet and Lee (1975) suggested using u/B<2/3 as the criteria to exclude the possibility of this kind of failure. The results shown in Fig. 2(a) indicate that the critical u/B value is a function of the effective friction angle of the soil. The criteria u/B<2/3 is valid for reinforced fill with the effective friction angle greater than 30^0.

Failure between reinforcements

The failure of the soil between reinforcements likely occurs for the situation involving a large spacing of reinforcements. As shown in Fig. 1(b), the ultimate bearing capacity for this kind of failure depends on: (1) shear resistance along the sides of the block above the uppermost reinforcement; (2) level of tension in the reinforcements; and (3) bearing capacity of the soil underneath the uppermost reinforcement. The level of tension in the reinforcements is attributed to the tensile strength at which overstressing occurs or the shear strength at large slippage between the reinforcements and the soil. Due to low overburden stresses for most model tests, slippage is most likely a controlling factor rather than the rupture or overstressing of the reinforcements. Guido, et al. (1987) reported that "in all 80 plate loading tests no sheet of geogrid was ever found to have ruptured during testing."

Deep punching failure

Deep punching failure, as depicted in Figure 1(c) commonly occurs when the underlying native soil is relatively soft. It is also likely to happen in reinforced foundations. The ultimate bearing capacity due to punching failure within a geosynthetic reinforced foundation can be estimated by modifying Meyerhof and

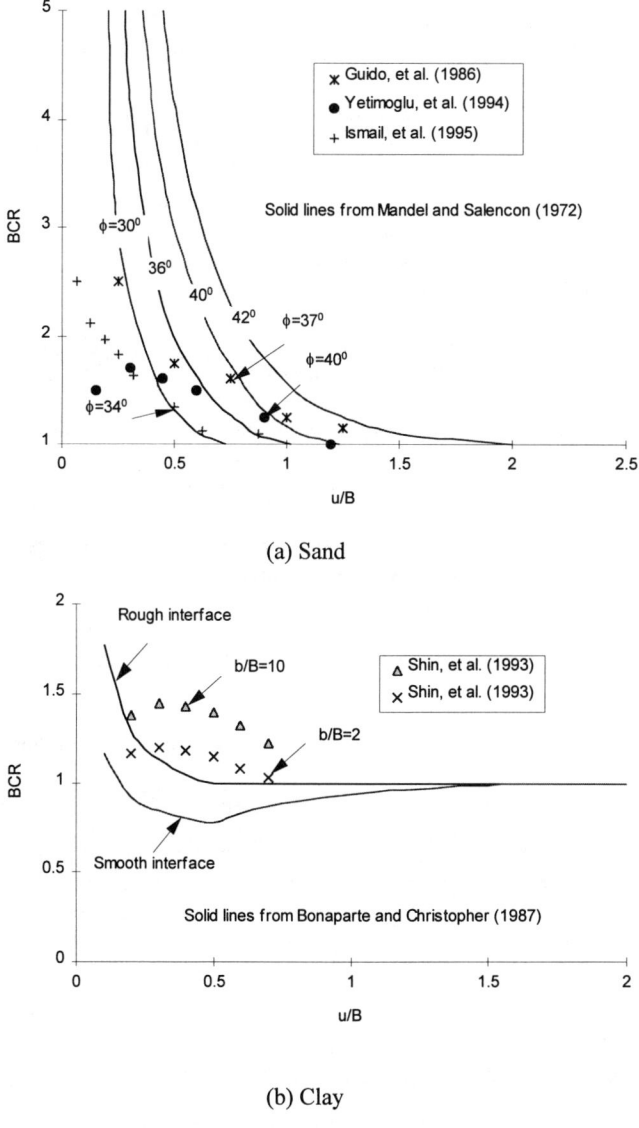

(a) Sand

(b) Clay

Figure 2. Bearing Capacity Ratio Due to Failure Above the Uppermost Reinforcement

Hanna's solution (Meyerhof and Hanna, 1978). The contribution of reinforcements to bearing capacity can be approximately modeled by an uplift or restraining force, T.

$$q_{ult} = q_b + 2c_a(B+L)\frac{H}{BL} + \gamma_1 H^2\left[1+2\frac{D}{H}\right]K_s(B+L)\frac{\tan\phi_1}{BL} + 2(B+L)\frac{T}{BL} - \gamma_1 H \quad (1)$$

where q_b = ultimate bearing capacity of the foundation below the reinforced zone;
c_a = unit adhesion of the upper layer;
γ_1 = unit weight of the upper layer;
H = thickness of the upper layer;
B = width of the footing;
L = length of the footing;
D = embedment depth of the footing;
ϕ_1 = friction angle of the upper layer;
K_s = punch shear coefficient for upper layer, which is dependent on ultimate bearing capacities of the upper and the lower soil layers;
T = uplift or restraining force of the reinforcements.

The restraining force, T, should be chosen as the lesser of the overstress strength of the reinforcement or the shear strength at large slippage. Based on the results obtained by Adams and Collin (1997) a comparison between the measured and the calculated bearing capacities has been performed. This information is presented in Table 1. The results are in good agreement with the values predicted by use of this equation.

Table 1 Measured and Calculated Bearing Capacities of Reinforced Foundations

B (m)	u/B	Reinforcement area (m^2)	Measured[1] q_{ult} (kPa)	Predicted q_{ult} (kPa)	Predicted Measured
0.61	0.25	1.2x1.2	360	359	1.00
0.61	0.38	1.2x1.2	370	425	1.15
0.61	0.25	1.8x1.8	360	365	1.01
0.61	0.38	1.8x1.8	398	435	1.09
0.61	0.25	2.4x2.4	360	372	1.03

[1] The measured ultimate bearing capacities were adopted from Adams and Collin (1997) at s/B=0.1. s is the settlement of the loading plate. The friction angle of the soil was back-calculated from loading test results for the unreinforced case.

Geosynthetic Reinforced Foundation (GRF) punching failure

The GRF punching failure mode is depicted in Figure 1(d). This situation can occur when the underlying soil is very soft, the reinforced mass is very strong, and the reinforced zone is not wide nor thick enough to reduce the stress at the base of the reinforced zone. The reinforced zone works as a rigid footing to punch into the underlying soft soil.

Critical failure mode

Although several potential failure mechanisms exist for a reinforced foundation, the actual failure of the foundation is always controlled by a critical mode. As such, the ultimate bearing capacity due to the critical failure mode is smaller than that due to any of the other modes of failure. The critical failure mode depends on the reinforcement configuration (distance of the reinforcement to the bottom of the footing, reinforced zone thickness, width and length of the reinforcement and spacing of subsequent layers) and the properties of the reinforcement and soil, and the pullout interaction coefficient between soil and reinforcement. As mentioned previously, the failure above the uppermost reinforcement may be the critical mode when the distance to the uppermost reinforcement is too large. The punching failure may be the controlling mode for the case when the reinforced zone is underlain by a very soft layer of soil. Each of these modes are evaluated in the DSS (1998) commercially available computer program that was developed by the authors.

Interpretation of Stress Distribution Angle from Experimental Data

Research work conducted by Gabr and Hart (1996b) demonstrated that a stress distribution angle of up to 45 degrees can be achieved with punched and drawn geogrids. This was verified by a field study performed by Wayne (1997) which involved the placement of 0.45 m of well graded sand over a plastic culvert. The punched and drawn geogrid with a junction strength at 2% strain of 3.4 kN/m in the machine direction and 4.4 kN/m in the cross machine direction was placed at a u/B ratio of 0.67. Subsequently, surface loads of up to 514 kPa were induced by incrementally loading the back end of a dump truck. In this study stresses were measured using the same pressure sensing devices as described by Gabr and Hart (1996b) which were placed at the interface between the plastic culvert and sand layer. Stresses for reinforced and unreinforced layers were measured. In Table 2 we can compare the measured values with those predicted by use of the following equation:

$$P_{os} = [(P_s/2)/[(B + 2h_{os}\tan \alpha)(L + 2h_{os} \tan \alpha)] + \gamma h_{os} \quad (2)$$

where: P_{os} = Pressure exerted at the soil/plastic culvert interface;
P_s = Axle load;
B = Tire contact length (0.3 m);
L = Tire contact width (0.45 m);

h_{os} = Soil layer depth (0.45 m for this study);
α = Stress distribution angle (a value of 26.6° was used to predict the stress for the unreinforced case and a value of 45° was used to predict the stress at the interface for the reinforced case in Table 2);
γ = Unit weight of the soil layer (18 kN/m³ for this study).

Table 2 - Measured Stresses Versus Calculated Stresses Based on Equation 2

Surface Stress (kPa)	392			514		
Stress at Depth (kPa)	Measured (M)	Predicted (P)	M/P	Measured (M)	Predicted (P)	M/P
Unreinforced	88.1	86.4	1.02	112.2	111.1	1.01
Reinforced	43.6	40.6	1.07	49.7	50.8	0.98
Reduction(%)	49.5	47.0	N/A	44.2	45.7	N/A

In lieu of using a predetermined stress distribution angle, we can back calculate the distribution angles using equation 2. For the 514 kPa surface stress the stress distribution angle was calculated as 26.3 degrees for the unreinforced case and 45.5 degrees for the reinforced case. For the 392 kPa surface stress the stress distribution angle was calculated as 26.1 degrees for the unreinforced case and 43.2 degrees for the reinforced case. And lastly, for a surface loading of 288 kPa, the stress distribution angle was calculated as 23.2 degrees for the unreinforced case and 46.4 degrees for the reinforced case. The area of loading used in these calculations was maintained as a constant value for the cases listed based on the basis of visual observations during the test program.

Settlement Design Issues

When any ground modification technique is considered as an alternate to deep foundations the main concern is settlement. In addition to figuring out how long the settlement process will take to complete (i.e., elastic (immediate) and/or consolidation settlement,) the geotechnical engineer is often asked to predict the amount and type of settlement that a particular site will experience under the influence of the proposed building(s). In particular, the structural engineer is concerned about the magnitude and mode of settlement in relation to the allowable settlement that the building(s) can withstand. The modes of settlement include (1) uniform settlement, (2) tilt, and (3) differential settlement. In the case of uniform settlement there is little concern about the structure and greater concern about entrance ways, and utilities. Tilt or angular distortion is measured as the difference in expected settlement over the length or width of the building [(S_{max} - S_{min}) / L] and the amount that can be tolerated by the building is a function of numerous factors. Tilt is visible at 1/250 and cracking due to angular distortion can occur at 1/300. Bowles (1982) indicates that the greatest differential settlement that is tolerable for buildings is 37.5 mm in clays and 25 mm in sands. Further, Bowles

(1982) recommends that maximum settlements for isolated foundations on or in clays not exceed 62.5 mm and 37.5 mm in sands. For raft foundations this recommendation is increased to 100 mm for buildings founded on clays and 62.5 mm for buildings constructed on sands. The sections which follow will discuss how the GRF is able to maintain differential settlement within tolerable limits and lead to a more accurate prediction of total settlement.

A key issue associated with settlement for GRF is the degree to which building loads are transferred to the GRF and the soil underlying the GRF. The reduction in total settlement is the result of (1) the geosynthetic reinforced foundation being stiffer than the in-situ soil, and (2) the fact that the vertical stresses induced at the interface between the in-situ soil and the geosynthetic reinforced foundation are less than that of the unreinforced soil. This is a result of the geosynthetic reinforced foundation distributing the spread footing load within the dimensions of reinforced zone. There are three settlement calculation methods incorporated within the DSS (1998) computer algorithm: (1) elastic theory, as presented by Lawton (1995), (2) a modified form of the Schmertmann (1970) and Schmertmann, et al. (1978), and (3) the stress distribution method.

Elastic theory

The situation involving the placement of a GRF over an unreinforced soil layer can be considered as a two layer system with a stiff layer overlying a relatively soft layer. The solutions for a two layer system as developed by Burmister (1962) is only applicable for layers with infinite width. As reported by Lawton (1995), Burmister's solutions may be useful if a considerable amount of lateral space is available for overexcavation/replacement over which foundation stresses must be transferred in order to achieve the full reduction in settlement. However, it is generally not economical when the overexcavation/ replacement zone extends too far beyond the boundary of the building foundation. The design charts developed by Lawton (1995) using a finite element method are more appropriate for overexcavation/replacement with a limited width including GRF's. This is because Lawton's design charts take into consideration a replaced zone with limited width and the influence of footing embedment. To compute the settlement of a footing on a replaced foundation, the following elastic solution can be used:

$$S_i = \frac{q_0 d}{E_2} I_s \qquad (3)$$

where, q_0 = uniform stress applied onto the surface of the replaced zone;
d = diameter of the loaded area;
E_2 = elastic modulus for the lower layer, and;

I_s = influence factor for settlement beneath the center of the loaded area.

As developed by Lawton (1995), the influence factor, I_s, is a function of four factors as shown in the following equation:

$$I_s = f(E_1/E_2, b/d, D/d, z/d) \quad (4)$$

in which E_1 is the elastic modulus of the replaced zone.

For GRF's, E_1 is the equivalent elastic modulus of the composite GRF zone. The back-calculated elastic modulus ratio, E_1/E_2, is based on this method and the experimental data collected by Adams and Collin (1997) as listed in Table 3. For reinforced foundations with two geogrid layers, the ratio ranges from 1.2 to 4.4, while the ratio for three geogrid layers ranges from 4.4 to 39.1. The variation of the modulus ratio is quite significant for the three geogrid layers. The reason for this variation needs to be investigated in the future. It should be pointed out that these ratios are only based on uniform cohesionless soils with punched and drawn geogrid inclusions. In practice, the reinforced zone is always compacted and the underlying soil is weak or soft. Therefore, the ratio will be even more significant in the field.

In terms of the assumptions, this method can only be used for circular footings seated on a replaced zone underlain by uniform and infinite soil. When the length to width ratio of rectangular footings is less than 1.5, this method can still provide an approximate solution.

Table 3 Back-Calculated Elastic Modulus Ratio

B (m)	0.61	0.61	0.61	0.61	0.61	0.61	0.46	0.61	0.61	0.91
N	2	2	2	2	2	2	3	3	3	3
u/B	0.41	0.41	0.41	0.41	0.41	0.41	0.72	0.41	0.41	0.19
z/B	1.02	1.02	1.02	1.02	1.02	1.02	1.17	0.88	0.88	0.58
γ (kN/m^2)	14.5	14.5	14.5	14.2	14.2	14.2	14.8	14.8	14.8	14.8
E_1/E_2	2.24	2.82	4.35	1.20	1.30	2.67	31.00	4.41	8.29	39.13
Average	2.43						20.7			

Modified Schmertmann method

The Schmertmann method (Schmertmann, 1970 and Schmertmann, et al., 1978) is modified for the case of GRF's to account for the stress distribution effect associated with this system. Based on the information presented earlier, the load applied onto the top of an overexcavated and replaced zone is transmitted to the bottom of this zone at a stress distribution angle which can range from 26.6^0 to 45^0.

The transmitted loading area at the bottom of the replaced zone is considered as the size of a new "footing". Below this new "footing", the Schmertmann method is adopted. Since the Schmertmann method was mainly developed for cohesionless soils, the modified method can only be applied to cohesionless soils.

Stress distribution method

As discussed earlier, a geosynthetic reinforced foundation can distribute applied loads from the top to the bottom of the reinforced foundation within a wider area compared with an unreinforced foundation. To obtain a reasonable estimate of the approximate slope of the vertical stress distribution [α(vertical):1(horizontal)] from a theoretical point of view, for the traditional overexcavation/replacement Lawton (1995), numerically integrated the curves developed by Burmister (1958) and Fox (1948) for a two layer elastic system as a function of the modulus ratio for the replaced layer to that of the in-situ soil layer or E_1/E_2. The results of this work are depicted in Figure 3. This, however, is for an unreinforced replaced zone and must be modified for the case of a geosynthetic reinforced foundation by considering an increase in the modulus ratio. The reader must keep in mind the fact that the replaced zone must extend beyond the boundaries of the stress distribution as determined in Figure 3. As Lawton (1995) also pointed out, the use of Figure 3 for sizing a replaced zone for rectangular or circular footings would often produce a required replaced zone that is much wider than the commonly used B to 3B range. Based on the interpretation of stress distribution angle from experimental data, as presented earlier, it is more reasonable to assume the stress distribution angle within the reinforced zone ranging from 26.6^0 to 45^0 (i.e. α = 2 to 1), which depends on width and behavior of the reinforced zone. As usual, the stress distribution angle within the unreinforced zone can be assumed to be 26.6^0. Therefore, an equivalent uniform stress distributed to a certain depth, H, can be determined through use of the following equations:

For rectangular foundations:

$$B_z = B + (2H/\alpha) \quad \text{(5)a}$$
$$L_z = L + (2H/\alpha) \quad \text{(5)b}$$
$$q_z = q_0 \frac{BL}{B_z L_z} \quad \text{(5)c}$$

For circular foundations:

$$d_z = d + (2H/\alpha) \quad \text{(6)a}$$
$$q_z = q_0 \frac{d^2}{d_z^2} \quad \text{(6)b}$$

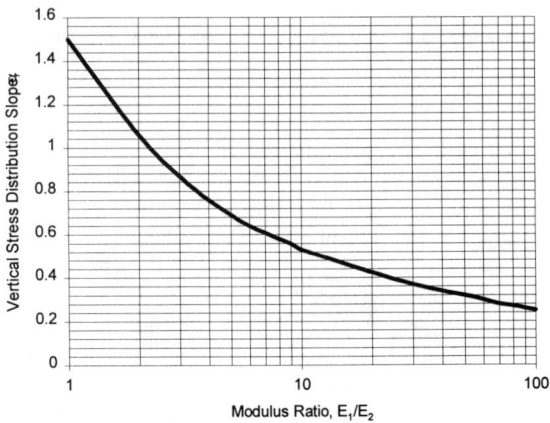

Figure 3. Vertical Stress Distribution for a Two-Layered Elastic System
(Lawton, 1995)

After computing the stresses at different depths, the settlement can be computed by accumulating the compression deformations with depth. This method can be used for cohesionless and cohesive soils. For cohesive soils, the compressive index should be determined while, and for cohesionless soils the elastic modulus should be estimated.

Typical Design Parameters for Geogrid Reinforced Foundations

Based on a comprehensive study of experimental data, literature review, and economical considerations, typical design parameters for punched and drawn biaxial geogrid reinforced foundations are recommended in Table 4 in order to obtain a bearing capacity ratio (BCR) of 1.5 to 2.5 for typical footing sizes in practice, such as 0.6 to 1.0 m wide strip footings and 1.5 to 1.8 m square footings. These typical design parameters are established based on the use of granular backfill materials for reinforced fill. In order to have good interaction between the lowest geogrid and fill, minimum 0.10 m thick backfill material should be placed below the lowest geogrid layer. In addition to bearing capacity, requirements for the settlement of footings also needs to be addressed.

Table 4 - Typical Design Parameters for Punched and Drawn Biaxial Geogrids

	Typical value	Recommended (not greater than)
u	0.15B to 0.3B	0.5B
s	0.15B to 0.3B	0.5B
z	0.5B to 1.0B	2.0B
b	2.0B to 3.0B	4.0B
a	0.1B to 0.2B	0.3B
Δl	0.5B to 1.0B	2.0B
N	2 to 4	5

Note:
B - footing width; u - distance from footing base to uppermost geogrid; s - spacing between geogrid layers; z - thickness of reinforced fill; b - width of reinforced fill; a - distance from lowest geogrid to bottom of reinforced fill; Δl - length of geogrid extended beyond each end of strip footing; N - number of geogrid layers.

Summary

After reviewing the vast amount of literature there is little doubt as to whether the geosynthetic reinforced foundations are a viable solution for improving bearing capacity or limiting total and differential settlements. In fact, laboratory and field tests have shown that this system can be used either in lieu of or in combination with other ground improvement techniques. This paper outlines controlling factors for bearing capacity and settlement of geosynthetic reinforced foundations. It also describes design methods which may be adopted for estimating bearing capacity and settlement of geosynthetic reinforced foundations. As a result of the work performed to date, the author's have developed commercially available analysis software that is capable of determining the bearing capacity and settlement for both unreinforced and geosynthetic reinforced foundations.

References

Adams, M.T. And Collin, J. G. (1997) "Large Model Spread Footing Load Tests on Geosynthetic Reinforced Soil Foundations," *Journal of Geotechnical and Geoenvironmental Engineering,* Vol. 123, No. 1, January 1997, pp. 66 - 72.

Akinmusuru and Akinbolade (1981), "Stability of Loaded Footings on Reinforced Soil," ASCE, *Journal of Geotechnical Engineering,* Vol. 107, No. GT6, pp. 819-827.

Bonaparte, R. and Christopher, B.R. (1987). "Design and Construction of Reinforced Embankments over Weak Foundation," Transportation Research Record 1153, pp. 25-39.

Burmister (1958), "Evaluation of Pavement Systems of the WASHO Road Test by Layered System Methods," Highway Research Board Bulletin 177, Washington, D.C., pp. 26-54.

Burmister (1962), "Application of Layered System Concepts and Principles to Interpretation and Evaluations of Asphalt Pavement Performances and to Design and Construction," Proceedings of the International Conference on Structural Design of Pavements, University of Michigan, Ann Arbor, pp. 441 - 453.

Chadbourne, W. (1994),"An Investigation Into The Performance of Shallow Spread Footing in Reinforced Cohesionless Soil," M.S. Thesis, Department of Civil and Environmental Engineering, Tufts University, Boston, 189 pp.

Das, B.M. (1994) "Bearing Capacity of Shallow Spread Footings on Geogrid-Reinforced Soil," Report Prepared for Tensar Earth Technologies, Inc., 95 pp.

Das, B.M. (1984) Principles of Foundation Engineering, PWS Publishers, Boston, MA, pp. 108 - 110.

Das, B.M., and Omar, M.T. (1994) "The effects of Foundation Width on Model Tests for the Bearing Capacity of Sand with Geogrid Reinforcement," Geotechnical and Geological Engineering, Chapman & Hall, Vol. 12, pp. 133-141.

DSS (1998) "DIMENSION™ Solution Software©," Tensar Earth Technologies, Inc., Atlanta, GA.

Fox, L. (1948), "The Mean Elastic Settlement of a Uniformly Loaded Area at a Depth Below the Ground Surface," Proceedings of the Second International Conference on Soil Mechanics and Foundation Engineering, Rotterdam, Netherlands, Vol. 1, pp. 129-132.

Fragaszy, R.J., Lawton, E. (1985), "Bearing Capacity of Reinforced Sand Subgrades," Journal of Geotechnical Engineering, ASCE, Vol. 110, No. 10, Oct., pp.1500 - 1507.

Fukuda, N. (1987), "Foundation Improvement by Polymer Grid Reinforcement," 8th Asian Regional Conference of Soil Mechanics and Foundation Engineering, Kyoto.

Gabr, M.A., and Hart, J.H., (1996a) "Load-Deformation Characteristics of Geogrid-Reinforced Soil Using Plate Load Tests," Research Report Prepared for Tensar Earth Technologies, Department of Civil Engineering, West Virginia University, Morgantown, West Virginia, 25 pp.

Gabr, M.A., and Hart, J.H., (1996b) "Stress Distribution Characteristics of Geogrid-Reinforced Soil Using Plate Load Tests," Research Report Prepared for Tensar Earth Technologies, Department of Civil Engineering, West Virginia University, Morgantown, West Virginia.

Guido, V.A., Biesiadecki, G.L. and Sullivan, M.J. (1985), "Bearing Capacity of a Geotextile Reinforced Foundation," Proceedings, XI International Conference on Soil Mechanics and Foundation Engineering, 3, A.A. Balkema, The Netherlands, pp. 1777-1780.

Guido, V.A., Knueppel, J.D., Sweeney, M.A., (1987), "Plate Loading Tests on Geogrid Reinforced Earth Slabs," Proceedings of the Geosynthetics '87 Conference, New Orleans, Volume 1, pp. 216-225.

Guido, V.A., Chang, D.K., Sweeney, M.A., (1986), "Comparison of Geogrid and Geotextile Reinforced Earth Slabs," Canadian Geotechnical Journal, Vol. 3, No. 4, pp. 435-440.

Guido, V.A., Chang, D.K., Sweeney, M.A., (1985), "Bearing Capacity of Shallow Foundations Reinforced with Geogrids and Geotextiles," Proceedings of the Second Canadian Symposium on Geotextiles and Geomembranes, Edmonton, Alberta, pp. 71-77.

Hansen, J.B. (1957). Foundations of structures: general report, 4th ICSMFE, vol. 2, pp. 441-447.

Hansen, J.B. (1970). "A revised and extended formula for bearing capacity." Danish Geotechnical Institute Bulletin No. 28, Copenhagen, 21pp.

Ismail, I. and Raymond, G.P. (1995),"Geosynthetic Reinforcement of Granular Layered Soils," *Proceedings of Geosynthetics '95,* Industrial Fabrics Association International, Volume 1, pp. 317 - 330.

Lade, P.V., and Lee, K.L. (1976), "Engineering Properties of Soils," Report No. UCLA-ENG-7652, University of California at Los Angeles.

Lawton, E.C. (1995) "Section 5A: Nongrouting Techniques," Practical Foundation Engineering Handbook, Edited by Robert Wade Brown, McGraw-Hill, pp. 5.3-5.400.

Mandel and Salencon (1972). "Force portante d'un sol une assise rigide." Geotechnique, Vol. 22, pp. 79-93.

Meyerhof, G.G. (1951). "The ultimate bearing capacity of foundations." Geotechnique, Vol. 2, No. 4, pp. 301-331.

Meyerhof, G.G. (1963). "Some recent research on the bearing capacity of foundations." Canadian Geotechnical Journal, Vol. 1, No. 1, pp. 16-26.

Meyerhof, G.G. and Hanna, A.M. (1978). "Ultimate bearing capacity of foundation on layered soils under inclined load." Canadian Geotechnical Journal, Vol. 15, pp. 565-572.

Milligan, G.W. E., and Love, J.P., (1984) "Model Testing of Geogrids Under an Aggregate Layer on Soft Ground," Symposium on Polymer Grid Reinforcement in Civil Engineering, London, 11 pp.

Meyerhof, G.G., and Hanna, A.M. (1978), "Ultimate Bearing Capacity of Foundations on Layered Soils Under Inclined Load," Canadian Geotechnical Journal, Vol. 15, No. 4, pp. 565-572.

Munfakh, G.A. (1996) "Ground Improvement Engineering - The State of the Practice," CONEXPO-CON/AGG '96, 30 polypropylene.

Ochiai, H., Tsukamoto, Y., Hayashi, S., Ōtani, J., and Ju, J.W. (1994) "Supporting Capability of Geogrid Mattress Foundation," The Fifth International Conference on Geotextiles, Geomembranes, and Related Materials, IGS, Singapore, September 5-9, pp. 321 - 326.

Schmertmann, J.H. (1970),"Static Cone to Compute Static Settlement Over Sand," *Journal of the Soil Mechanics and Foundations Division,* American Society of Civil Engineers, Vol. 96, No. SM3.

Schmertmann, J.H., Hartman, J.P., and Brown, P.R. (1978),"Improved Strain Influence Factor Diagrams," *Journal of the Soil Mechanics and Foundations Division,* American Society of Civil Engineers, Vol. 104, No. GT8.

Terzaghi, K. (1943). Theoretical Soil Mechanics. John Wiley & Sons, Inc., New York, 510pp.

Tsukada, Y., Isoda, T., and Yamanouchi, T (1993) "Geogrid Subgrade Reinforcement and Deep Foundation Improvement: Yono City, Japan," Raymond, G.P. and Giroud, J.P. editors, International Society for Soil Mechanics and Foundation engineering, Committee TC9, Geosynthetics Case Histories, March, pp. 158 - 159

University of California (1984), "Plate Bearing Tests on Tensar SS1 Geogrid Reinforced Granular Base on Weak Clay Subgrade, Department of Civil Engineering Report to Tensar.

Valsangkar, A.J., and Meyerhof, G.G. (1979), "Experimental Study of Punching Coefficients and Shape Factor for Two-Layered Soils," Canadian Geotechnical Journal, Vol. 16, No. 4, November, pp. 802-805.

Wayne, M.H. (1997), "Field Study on the Performance of Tensar DIMENSION™ geogrid with plastic culverts," Internal Report, Tensar Earth Technologies, Inc. Atlanta, GA.

DYNAMIC LOADING ON FOUNDATION ON REINFORCED SAND

By B. M. Das[1], Member, ASCE

ABSTRACT: Laboratory model tests for the settlement of a surface square foundation supported by a medium dense sand and subjected to cyclic loading of low frequency (1 cps), and transient loading have been presented. These tests were conducted with and without geogrid reinforcement in the soil. Only one type of soil and one type of geogrid were used for all tests. The maximum permanent settlement due to the cyclic and transient loads in reinforced and unreinforced soils have been compared. Based on the present tests, it appears that geogrid reinforcement can act as a settlement retardant for dynamic loading conditions on the foundations.

INTRODUCTION

Several studies are now in progress to evaluate the effectiveness of using geogrid as reinforcement to increase the ultimate and allowable load bearing capacities of shallow foundations in various types of soil (e.g., Guido et al., 1986; Omar et al., 1993; Adams et al., 1997). The majority of the published studies are based on small scale model tests conducted in the laboratory. Only a few studies such as Adams et al. (1997) provide full-scale field test results.

In some cases shallow foundations which are supported by geogrid-reinforced soil may be subjected to *cyclic loading* and *transient loading*. These problems will primarily be encountered in vibratory machine foundations. This paper presents the results of some laboratory model tests related to the permanent settlement of a *square surface foundation* supported by geogrid-reinforced sand and subjected to cyclic and transient loading conditions.

PROBLEM STATEMENT

Figure 1 shows a square surface foundation measuring $B \times B$ in plan on a sand layer which is reinforced with N layers of geogrid. The dimensions of the geogrid layers are

----

[1] School of Engineering and Computer Science, California State University, Sacramento, CA 95819-6023.

$b \times b$. The first layer of geogrid is located at a distance u from the bottom of the foundation. The distance between each consecutive layer of geogrid is h. Referring to Fig.1, the distance d measured from the bottom of the foundation to the bottom layer of geogrid can be given as

$$d = u + (N - 1)h \tag{1}$$

Past model test results have shown that the ratio of u/B should be kept less than about 0.67 to 1 in order to derive most of the beneficial effects of soil reinforcement. It has also been shown that, for given values of u/B and h/B, there are maximum values of u/B and d/B beyond which the ultimate bearing capacity remains practically constant or increases by a very small amount. Hence, for all practical purposes, these maximum values may be considered to be the critical values $(b/B)_{cr}$ and $(d/B)_{cr}$.

For any given combination of u/B, h/B, b/B and d/B, if the foundation is subjected to an allowable static load of intensity q_s, it will undergo an elastic settlement of S_s. Now, if a cyclic load of intensity q_{dc} having an amplitude of $q_{dc(max)}$ is superimposed on the static load, the nature of foundation settlement will be of the type shown in Fig.2. The maximum settlement due to the cyclic load will be $S_{dc(max)}$ which will be a function of q_s, $q_{dc(max)}$, stiffness of the geogrid, depth of reinforcement, number of geogrid layers, and other soil parameters. For similar soil conditions, q_s and $q_{dc(max)}$, the presence of the geogrid reinforcement layers will have an effect in reducing the magnitude of $S_{dc(max)}$ when compared to that in unreinforced soil.

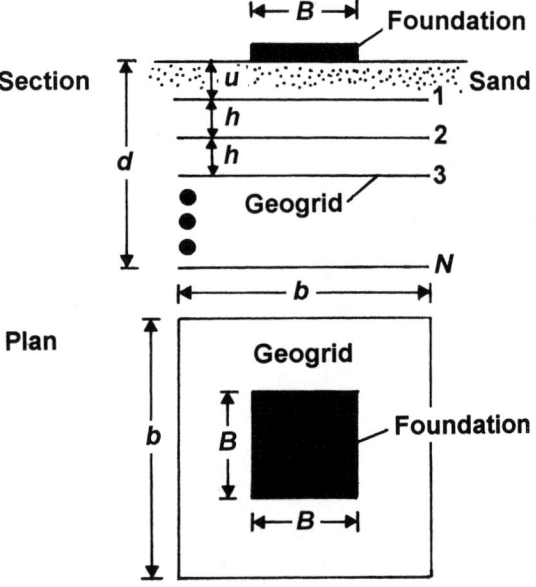

FIG. 1. Square Surface Foundation on Geogrid-Reinforced Sand

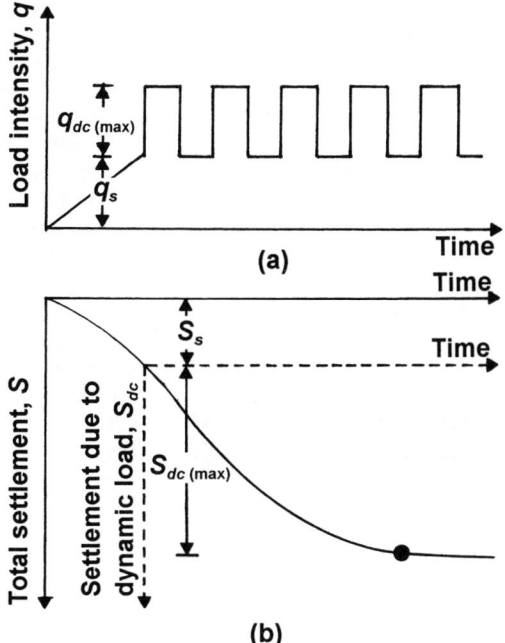

FIG. 2. Nature of Variation of (a) Load Intensity With Time; (b) Settlement

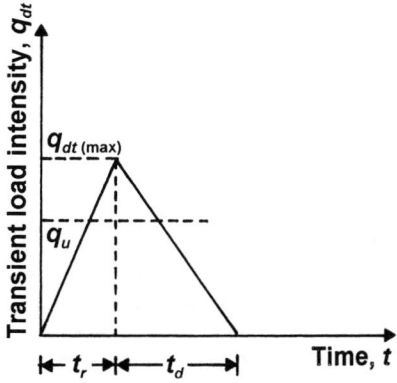

FIG. 3. Transient Load on a Foundation

It is also likely that machine foundations, in some instances, may be subjected to occasional transient loading of short duration where the maximum intensity of load [$q_{dt\,(max)}$] may exceed the *ultimate bearing capacity of unreinforced soil* (q_u). The loading condition being referred to is shown in Fig. 3. Note that the transient load has a rise time of t_r and decay time of t_d. Depending on the magnitude of $q_{dt\,(max)}$ and q_u, the foundation may undergo large settlement (S_{dt}). One possible way to reduce the undesirable settlement is to use geogrid as reinforcement in the soil supporting the foundation.

In order to quantify some of the parameters which control the settlement of the foundation due to the dynamic loading conditions described above, three series of laboratory model tests in sand were conducted for this study. They are as follows:

- Series A: Bearing capacity tests with and without geogrid reinforcement to determine $(b/B)_{cr}$ and $(d/B)_{cr}$.
- Series B: Cyclic load tests on unreinforced sand and geogrid-reinforced sand [with $b/B = (b/B)_{cr}$ and $d/B = (d/B)_{cr}$] with varying magnitudes of q_s and $q_{dc\,(max)}$ to determine $S_{dc(max)}$.
- Series C: Tests with transient load on the model foundation supported by sand with and without geogrid reinforcement. The magnitude of $q_{dt\,(max)}/q_u$ was varied. The magnitude of settlement due to the transient load (S_{dt}) with time was monitored.

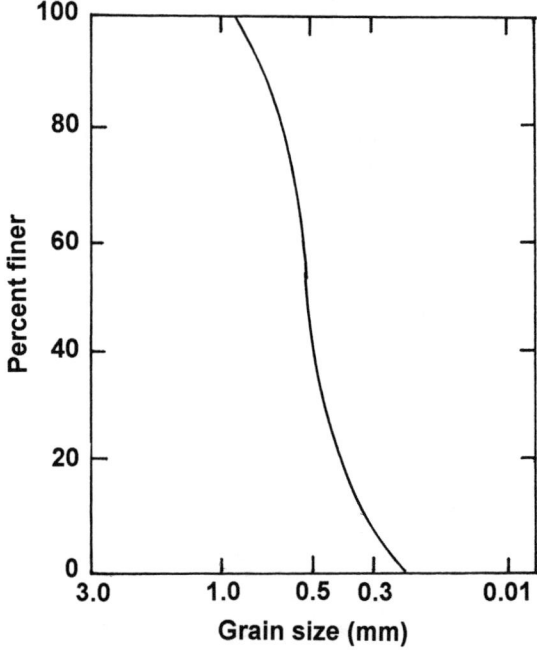

FIG. 4. Grain-Size Distribution of Sand Used for the Model Tests

MATERIAL PROPERTIES AND MODEL TEST ARRANGEMENT

Laboratory tests were conducted using a model foundation made of aluminum and measuring 76.2 mm × 76.2 mm. The bottom of the model foundation was made rough by cementing a thin layer of sand to it. The model tests were conducted in a box measuring 760 mm × 760 mm × 760 mm. A fine silica sand was used for the present tests. Figure 4 shows the grain-size distribution of the sand. A biaxial geogrid (TENSAR BX1000) was used for soil reinforcement. The properties of the geogrid are given in Table 1.

TABLE 1. Properties of the Geogrid

Item (1)	Property (2)
Structure	Punched sheet drawn
Polymer	Polypropylene/High density polyethylene co-polymer
Junction method	Unitized
Aperture size (MD/XMD)	25.4 mm/33.02 mm
Nominal rib thickness	0.75 mm
Nominal junction thickness	2.29 mm
*Peak tensile strength (MD)	8.46 kN/m
*Tensile strength @ 2% strain (MD)	2.33 kN/m
*Peak tensile strength (XMD)	11.97 kN/m
*Tensile strength @ 2% strain (XMD)	2.92 kN/m
*Initial tangent modulus (MD)	218.9 kN/m
*Initial tangent modulus (XMD)	364.8 kN/m

*Based on manufacturer's specifications

In order to conduct a model test, sand was poured into the test box in 25.4-mm thick layers, up to the desired height, by using a raining technique. Geogrid reinforcements, when needed, were placed in the sand at predetermined values of u/B and h/B. The average unit weight of compaction and the relative density of compaction of sand were 17.4 kN/m^3 and 76%, respectively. The average peak friction angle at that average relative density of compaction was determined by direct shear test to be about 42°.

BEARING CAPACITY TESTS (SERIES A)

As mentioned before, the tests in this series were conducted to determine the critical reinforcement-depth ratio $[(d/B)_{cr}]$ and the critical reinforcement-width ratio $[(b/B)_{cr}]$. In conducting the tests, load to the model foundation was applied by a hydraulic jack. The loading ram was fixed to the model foundation and the foundation was not allowed to rotate. The settlement corresponding to a given load was measured by two dial gauges. The details of the tests conducted in this series are given in Table 2.

TABLE 2. Details of Bearing Capacity Tests (Series A)

Test series (1)	Constant parameters (2)	Variable Parameters (3)	Comments (4)
A – 1	$D_r = 76\%$	— — —	To determine the ultimate bearing capacity, q_u, on unreinforced sand
A – 2	$D_r = 76\%$ $u/B = h/B = 1/3$ $b/B = 6$	$N = 1, 2, 3,$ $4, 5, 6$	To determine the ultimate bearing capacity, $q_{u(R)}$, on reinforced sand and thus $(d/B)_{cr}$
A – 3	$D_r = 76\%$ $u/B = h/B = 1/3$ $N = 4$	$b/B = 1, 2, 3,$ $4, 5, 6$	To determine the ultimate bearing capacity, $q_{u(R)}$, on reinforced sand and thus $(b/B)_{cr}$

D_r = relative density

All of the load-settlement curves showed a peak value which was the ultimate bearing capacity. Figure 5 shows the ultimate bearing capacities q_u (on unreinforced sand) and $q_{u(R)}$ (on reinforced sand) as obtained from Series A–1 and A–2. The magnitude of $S/B = S_u/B$ (S = settlement; S_u = settlement at ultimate load) at ultimate load was about 3.5% for tests in unreinforced sand (Series A–1). For tests in Series A–2 conducted on reinforced sand, the magnitude of S_u/B increased from about 5% for $N = 1$ to about 6.5% for $N = 6$. From Fig. 5 it can be seen that the magnitude of the ultimate bearing capacity rapidly increased with d/B up to practically a maximum value at $d/B \approx 1.33$. However for $d/B >$ about 1.33, the slope of $\Delta q_{u(R)}/\Delta(d/B)$ reaches a minimum value. Hence, $d/B = 1.33$ may be taken to be approximately equal to $(d/B)_{cr}$.

The variation of ultimate bearing capacities q_u and $q_{u(R)}$ obtained from Test Series A–1 and A–3 is shown in Fig. 6. The magnitude of S_u/B for Series A-3 tests varied between 5% to about 6%. For tests conducted in Series A–3 with reinforced sand, the reinforcement depth ratio (d/B) was kept to be equal to $(d/B)_{cr} \approx 1.333$ as determined from tests in Series A–2. From Fig. 6 it can clearly be seen that for $b/B > 4$, the increase in the ultimate bearing capacity is minimal. Thus it can be assumed that $(b/B)_{cr} \approx 4$.

CYCLIC LOAD TESTS (SERIES B)

The cyclic load tests were conducted without (Series B–1, B–2, and B–3) and with (Series B–4, B–5, and B–6) soil reinforcement. For all tests conducted with soil reinforcement, the magnitudes of d/B and b/B were kept at 1.333 [$\approx (d/B)_{cr}$] and 4 [$\approx (b/B)_{cr}$], respectively. Also, as in Test Series A–2 and A–3, $u/B = h/B$ was kept equal to 1/3.

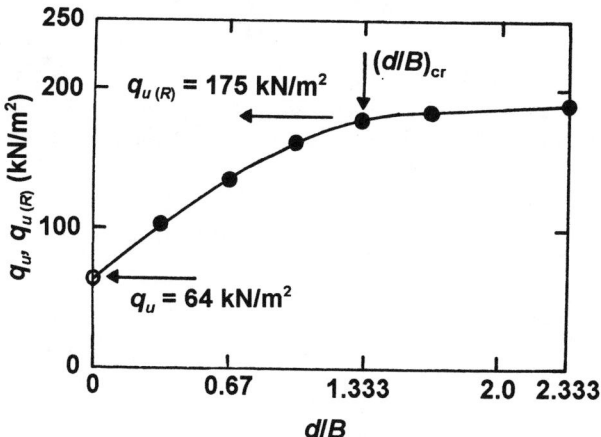

FIG. 5. Variation of q_u and $q_{u(R)}$ Obtained From Test Series A–1 and A–2 (*Note:* For Test Series A–2, $u/B = h/B = 0.333$ and $b/B = 6$)

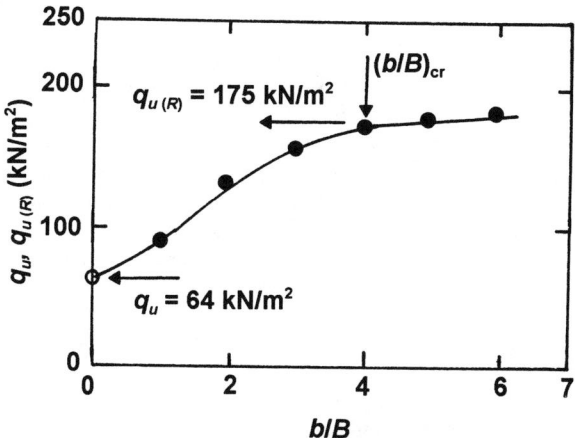

FIG. 6. Variation of q_u and $q_{u(R)}$ Obtained From Test Series A–1 and A–3 [*Note:* For Tests in Series A–3, $u/B = h/B = 0.333$ and $N = 4$, i.e., $d/B = (d/B)_{cr} = 1.333$]

For conducting the cyclic load tests, a Universal testing machine was used to apply load on the model foundation. The magnitude of the load and the foundation settlement were recorded by a data acquisition system. In order to start a test, a static allowable load per unit area, q_s, with a desired factor of safety ($FS = q_{u(R)}/q_s$) was first applied to the model foundation followed by the application of the cyclic load of intensity, q_{dc}. The period, T, of the cyclic load for all tests was 1.0 s. Details of the laboratory tests conducted in this phase of the study are given in Table 3.

TABLE 3. Details of Cyclic Load Tests (Series B)

Test series (1)	$FS = q_{u(R)}/q_s$ (2)	$q_{dc(max)}/q_{u(R)}$ (%) (3)	Remarks (4)
B – 1	7.6	4.36, 10.67 and 14.49	Tests with unreinforced sand
B – 2	4	4.36, 10.67 and 14.49	
B – 3	3	4.36, 10.67 and 14.49	
B – 4	7.6	4.36, 10.67, 14.49 and 22.33	Tests with soil reinforcement; $d/B = (d/B)_{cr} \approx 1.333$ $u/B = h/B = 0.333$ $b/B = (b/B)_{cr} \approx 4$
B – 5	4	4.36, 10.67, 14.49 and 22.33	
B – 6	3	4.36, 10.67, 14.49 and 22.33	

Note: $q_{u(R)} \approx 175$ kN/m² as obtained from Series A tests with reinforcement

Figure 7 shows the results obtained from Test Series B–3 (unreinforced sand) and B–6 (tests with soil reinforcement) which is a plot of S_{dc}/B (S_{dc} = permanent settlement due to cyclic load application) versus number of load cycle applications, n, for various values of $q_{dc(max)}/q_{u(R)}$. Note that the magnitude of FS for B–3 and B–6 series is equal to 3. For any given test, the magnitude of S_{dc} increases with n and reaches practically a constant maximum value, $S_{dc(max)}$, at $n = n_{cr}$. Similar results were obtained from Test Series B–1, B–2, B–4, and B–5.

The maximum permanent settlement, $S_{dc(max)}$, obtained from all tests in Series B is shown in Fig. 8. Based on the Series B tests, the following general observations can be made:

1. For given values of FS and n, the magnitude of S_{dc}/B increases with the increase of $q_{dc(max)}/q_{u(R)}$.
2. If the magnitude of $q_{dc(max)}/q_{u(R)}$ and n remain constant, the value of S_{dc}/B increases with a decrease in FS.
3. The magnitude of n_{cr} for all tests in reinforced soil is approximately the same, varying between 1.75×10^5 and 2.5×10^5 cycles. Similarly, the magnitude of n_{cr} for all tests in unreinforced soil varies between 1.5×10^5 and 2.0×10^5 cycles.

Using the experimental values of $S_{dc(max)}$ given in Fig. 8, the settlement ratios $S_{dc(max)\text{-reinforced}}/S_{dc(max)\text{-unreinforced}}$ for various combinations of $q_{dc(max)}/q_{u(R)}$ and $FS = q_{u(R)}/q_s$ have been calculated and are shown in Fig. 9. From this figure it can be seen that, although there is some scatter, the settlement ratio varies linearly with $q_{dc\,(max)}/q_{u(R)}$ and is not dependent on *FS*.

TRANSIENT LOAD TESTS (SERIES C)

This series consisted of tests in which the model surface foundation was supported by unreinforced sand, and sand reinforced by layers of geogrid which were subjected to transient loads in which $q_{dt\,(max)}/q_{u(R)}$ (Fig. 3) was equal to and greater than one. Note that q_u is the ultimate bearing capacity in *unreinforced* soil. The transient load was applied using an MTS machine under its load control mode. For the transient loads the maximum value was reached at an average time $t_r = 1.75$ s and returned to zero at an average time $t_d = 1.4$ s (variation in t_r and t_d was about ±10%). The loading data was collected through

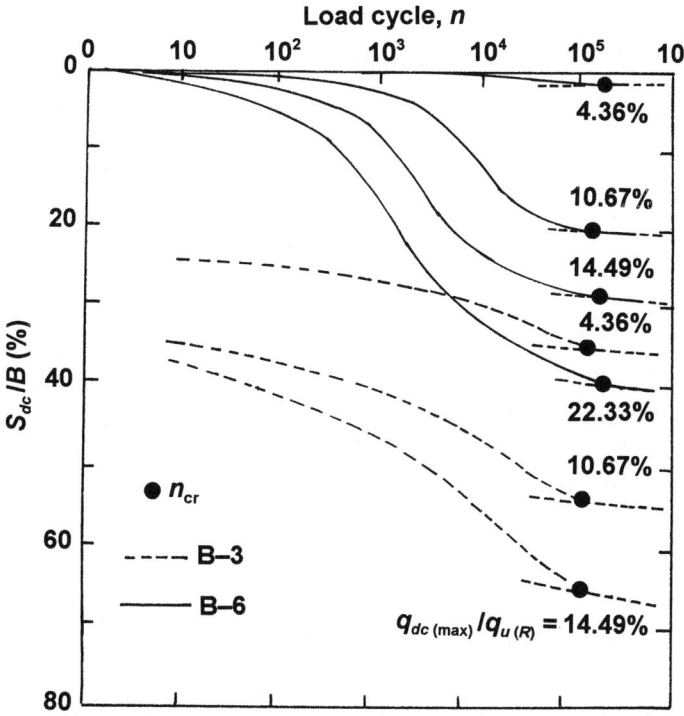

FIG. 7. Plot of S_{dc} Versus Number of Load Cycle Applications—Test Series B–3 (unreinforced sand) and B–6 (reinforced sand; $u/B = h/B = 0.333$, $b/B = 4$, $d/B = 1.333$)

the load cell of the machine. The foundation settlement was supported by two linear variable differential transducers. Each LVDT was mounted on an L-shaped aluminum arm rigidly attached to the outer wall of the test box. The analog signals of the LVDTs and load cell were recorded simultaneously by a data acquisition system. The data acquisition mechanism was triggered as soon as the loading started. Table 4 provides the details of the model tests. For all tests with reinforced soil, b/B was kept at 4 which was equal to $(b/B)_{cr}$. Before the tests were started, it was expected that the transient loading pattern with $t_r = t_d = 1.5$s would be triangular as shown in Fig. 3. However, during the tests, the actual loading patterns obtained were somewhat irregular and unavoidable.

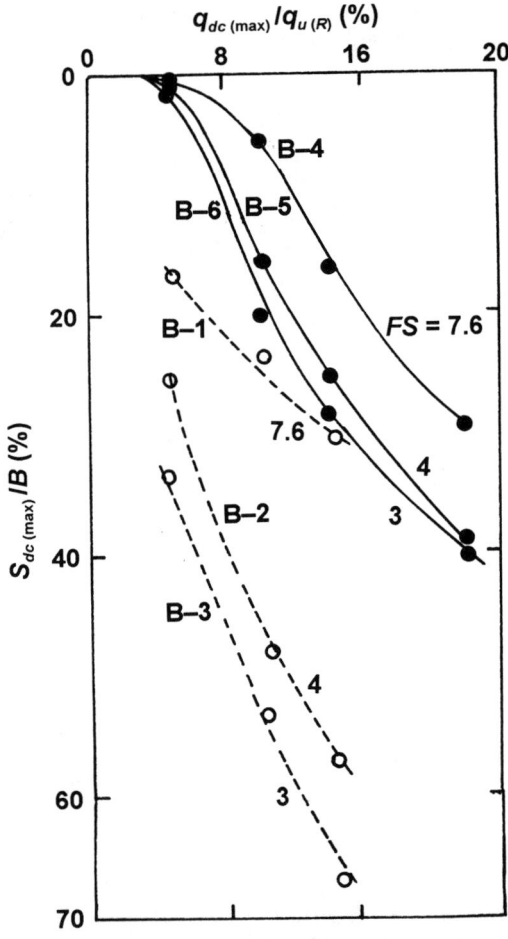

FIG. 8. Variation of $S_{dc(max)}/B$ With $q_{dc(max)}/q_{u(R)}$ and FS—Test Series B

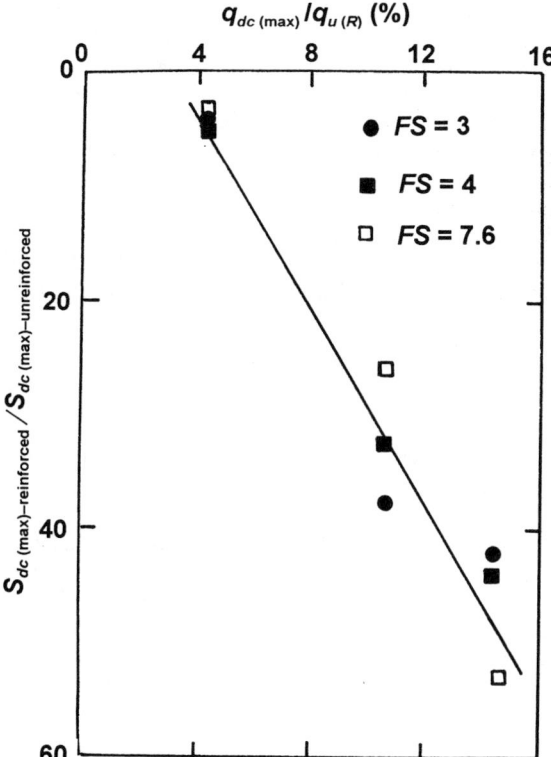

FIG. 9. Variation of Settlement Ratio–Test Series B

TABLE 4. Details of Transient Load Tests (Series C)

Series (1)	Reinforcement (2)	d/B (3)	b/B (4)
C – 1	Unreinforced sand; $N = 0$	— —	— —
C – 2	Geogrid-reinforced sand; $u/B = 1/3; N = 1$	1/3	4
C – 3	Geogrid-reinforced sand; $u/B = h/B = 1/3; N = 2$	2/3	4
C – 4	Geogrid-reinforced sand; $u/B = h/B = 1/3; N = 3$	1.0	4
C – 5	Geogrid-reinforced sand; $u/B = h/B = 1/3; N = 4$	1-1/3 $\approx (d/B)_{cr}$	4

Figure 10 shows typical plots of the variation of q_{dt} (transient load per unit area) and S_{dt} (settlement of foundation due to the transient load) with time. The variations of $S_{dt\,(max)}$

/S_u [$S_{dt\,(max)}$ = maximum settlement due to the transient load; S_u = settlement of the foundation at ultimate load when supported by *unreinforced* sand in Series A–1] with $q_{dt\,(max)}/q_u$ obtained from all the tests conducted in Series C are shown in Fig. 11. As expected, there was some scattering. Therefore, based on the average plots, the following general observations can be made.

1. For a given value of $q_{dt\,(max)}/q_u$, the settlement of the foundation decreased with the increase in the number of layers of geogrid.
2. For a given number of reinforcement layers, the magnitude of $S_{dt\,(max)}$ increased with the increase in $q_{dt\,(max)}/q_u$.

The effectiveness of geogrid reinforcement in reducing the settlement can be visualized by defining a quantity called the settlement reduction factor, R, or

$$R = \frac{S_{dt(max)-d}}{S_{dt(max)-d=0}} \qquad (2)$$

where $S_{dt\,(max)-d}$ = maximum settlement due to the transient load with reinforcement depth d, and $S_{dt(max)-d=0}$ = maximum settlement due to transient load with no soil reinforcement ($N = 0$).

Based on the results shown in Fig. 11 and Eq. (2), the settlement reduction factors have been calculated and plotted in Fig. 12. From this figure it can be seen that

1. For a given depth of reinforcement, the magnitude of R increases with the increase in $q_{dt(max)}/q_u$.
2. For a given value of $q_{dt\,(max)}/q_u$, the transient load-related settlement decreases rapidly with the increase in the depth of reinforcement. For the tests under consideration, the magnitude of R was about 0.15 for $q_{dt\,(max)}/q_u = 3$ when the depth of reinforcement was equal to d_{cr}. Hence, geogrid reinforcement is a settlement retardant for unanticipated transient loading on the foundation.

GENERAL COMMENTS

There are several shortcomings in the model tests that are presented in this paper, and they are as follows.

1. As in the case of practically all previously published model test results, full scale geogrid was used in the soil reinforcement for testing a model foundation. This introduces a mismatch of geometry, strength and, more importantly, stiffness between the model behavior and prototype response. In future studies, modeling of geogrid should also be addressed. However it needs to be mentioned that the geogrid used for this test program is one of the weakest available in the U.S. In practically all cases, stronger geogrid are used in the field.
2. The present test results are based on tests conducted on a small model foundation. Future tests need to be conducted with larger foundations [such as an extension of the study by Adams et al. (1997) with foundation subjected to dynamic loading] to validate the present findings and determine the existence of scale effect, if any.

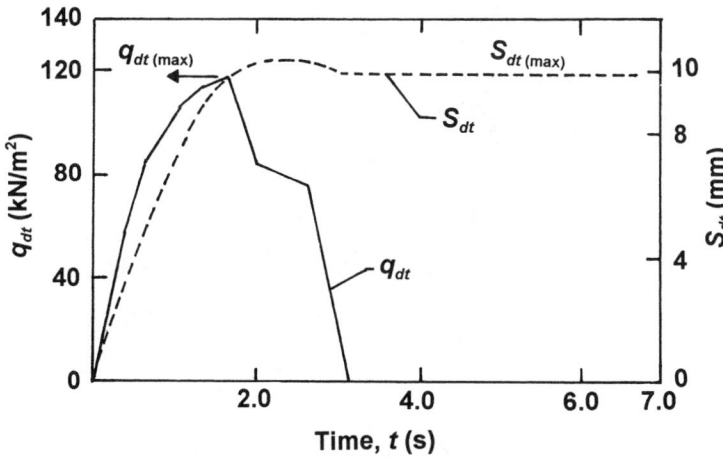

FIG. 10. Variation of q_{dt} and S_{dt} With Time—Test Series C-2 (Note: u/B = 0.333 and N = 1, i.e., d/B = 0.333)

CONCLUSIONS

Laboratory model test results for the permanent settlement of a square surface foundation supported by a medium dense sand and subjected to cyclic loads of low frequency, and transient loads have been presented. These tests were conducted in unreinforced soil and also in soil reinforced with layers of geogrid. Based on the model test results, the following conclusions can be drawn.

1. For realization of the maximum possible ultimate bearing capacity due to geogrid reinforcement, the critical depth of reinforcement is about $1.33B$, and the critical width of reinforcement layers is about $4B$.
2. Geogrid reinforcement has an influence in reducing the maximum permanent settlement of the foundation when subjected to cyclic loading. However the magnitude of the maximum permanent settlement is a function of q_s, $q_{dc\,(max)}$, relative density of the soil, and also the stiffness of the geogrid.
3. Geogrid reinforcement greatly reduces the permanent settlement of the foundation when subjected to transient loading. The magnitude of the maximum permanent settlement is a function of the depth of reinforcement, $q_{dc\,(max)}$, and q_u.

APPENDIX I. REFERENCES

Adams, M. T., Lutenegger, A. J., and Collins, J. G. (1997). "Design implications of reinforced soil foundation using soil strain signature and normalized settlement criteria," *Proc., Intl. Symposium on Mechanically Stabilized Backfill*, Denver, A. A. Balkema, 159-165.

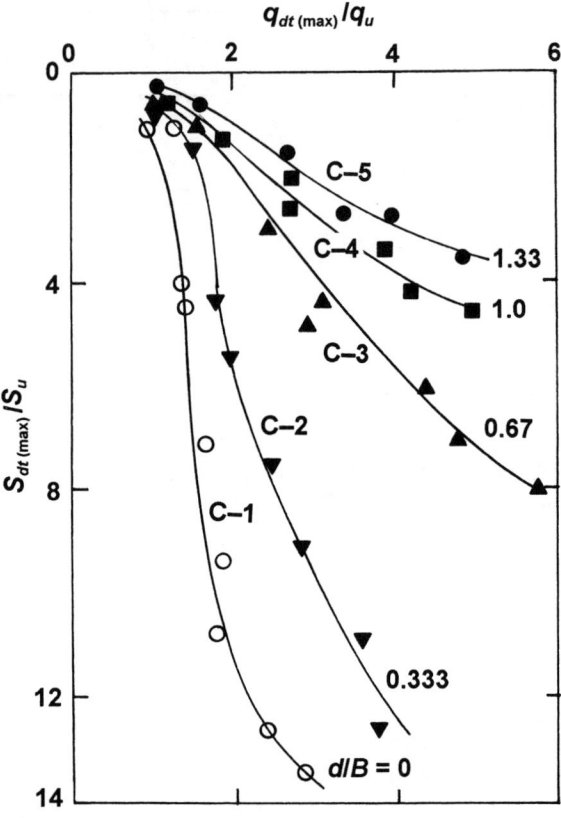

FIG. 11. Plot of $S_{dt(max)}/S_u$ Versus $q_{dt(max)}/q_u$—Test Series C

Guido, V. A., Chang, D. K., and Sweeny, M. A. (1986). "Comparison of geogrid and geotextile reinforced slab," *Canadian Geotech. J.*, 23(4), 435-440.

Omar, M. T., Das, B. M., Yen, S. C., Puri, V. K., and Cook, E. E. (1993). "Ultimate bearing capacity of rectangular foundation on geogrid-reinforced sand," *Geotech. Testing J.*, ASTM, 16(2), 246-252.

APPENDIX II. NOTATION

B = width of foundation
b = width of geogrid layers
d = depth of reinforcement
D_r = relative density

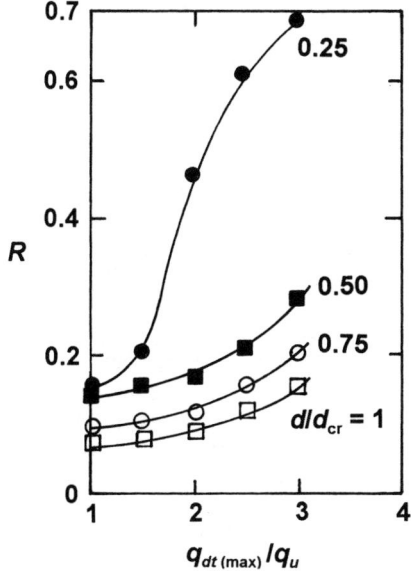

FIG. 12. Plot of Settlement Reduction Factor—Test Series C

FS = factor of safety = $q_{u(R)}/q_s$
h = distance between consecutive geogrid layers
N = number of geogrid layers
n = number of load cycle application
n_{cr} = critical number of load cycles
q_{dc} = intensity of cyclic load
$q_{dc(max)}$ = amplitude of cyclic load intensity
q_{dt} = intensity of transient load
$q_{dt(max)}$ = maximum intensity of transient load
q_s = allowable static load intensity
q_u, $q_{u(R)}$ = ultimate bearing capacity, respectively, in unreinforced and reinforced soil
R = settlement reduction factor
S = settlement
S_{dc} = settlement due to cyclic load
S_{dt} = settlement due to transient load
S_s = settlement due to static load intensity, q_s
S_u = settlement at ultimate load
T = period of cyclic load
t_r, t_d = rise and decay time of transient load, respectively
u = distance between the bottom of the foundation and first layer of geogrid

Behavior of the first prototype and full-scale models of PLPS geosynthetic-reinforced soil structure

Taro UCHIMURA[1], Fumio TATSUOKA[2], Masaru TATEYAMA[3], Tetsushi KOGA[4]

Abstract
It is shown that even very densely compacted reinforced backfill could exhibit large creep deformation by static loading and residual deformation by cyclic loading. A new construction method by means of vertical preloading and prestressing is described, which makes reinforced soil structures very stiff, exhibiting very small creep, transient, and residual deformations. The performance of a railway bridge pier, constructed in 1996 as the first preloaded and prestressed geosynthetic-reinforced soil structure, is presented, compared with the behavior of three full-scale models during loading tests.

Introduction

Geosynthetic-reinforced soil retaining walls (GRS-RW) with full height rigid facings have been constructed for a total length more than 25 km in Japan to support railway and highway (Tatsuoka et al. 1997b). Their popular use can be attributed to their high performance as well as a high performance/cost ratio. Some walls exhibited high seismic stability during the 1995 Hyogoken-nanbu earthquake. Some other applications have been bridge abutments to support relatively short girders. However, they have not been used for supporting heavier and more important structures, for example, abutments or bridge piers for long girders. The major reason is their relatively low stiffness against vertical compressive load applied on the top of backfill, compared to the conventional reinforced concrete (RC) abutments.

The authors have developed the preloaded and prestressed

---

1 and 2 : Professor and Research Associate, Department of Civil Engineering, University of Tokyo, 7-3-1, Bunkyo-ku, Hongo, Tokyo 113-8656 Japan; 3 : Senior Research Engineer, Railway Technical Research Institute; 4 : Engineer, Kyushu Railway Company

(PLPS) reinforced soil method since 1995 (Tatsuoka et al., 1997a). This method aims at substantially increasing the stiffness of geosynthetic-reinforced soil structures against short- and long-term loads by applying vertical preload (PL) and prestress (PS). A railway bridge pier as the first prototype PLPS reinforced soil structure was constructed, which was opened to service in the summer 1997. In parallel, three full-scale models of PLPS reinforced soil backfill were constructed to investigate the feasibility and characteristics. In this paper, the properties and the behaviors of the pier and the model embankments are compared to evaluate the effects of compaction, preload level, prestress level, and the amplitude of cyclic load on the structural behavior.

PLPS reinforced soil method

Two rigid reaction blocks, which are usually made of reinforced concrete, are placed at the top and bottom of the reinforced backfill (Fig.1). Four (or more) tie rods, for example steel rods as usually used for prestressed concrete, are vertically inserted through the backfill and connected to the reaction blocks. In order to make the reinforced soil mass behave elastically under working loads, sufficiently large preload is first applied by introducing tension into the tie rods using jacks placed at the top end of the tie rods. By being reinforced, the backfill soil can sustain very large preload without failure. Subsequently, the tie rod tension is reduced to a prescribed level, and the tie rods are connected to the top reaction block. Some tensile force T remains in the tie rods, with corresponding compressive load C remaining in the backfill soil, functioning as the prestress at this and subsequent stages. This prestress keeps high confining pressure to the backfill, which in turn achieves high soil stiffness. This procedure is based on the fact that the Young's modulus for vertical elastic strain increments of granular materials is a unique function of the current vertical normal stress (Tatsuoka and Kohata, 1995;

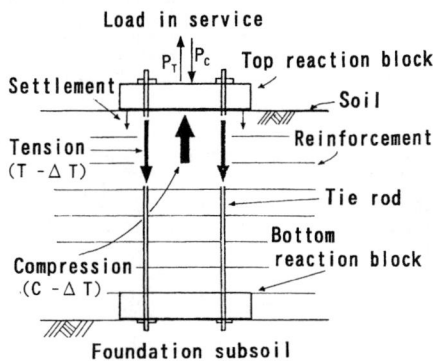

Fig. 1 Forces acting on PLPS reinforced soil mass

Tatsuoka et al., 1997a).

When either compressive load P_C or tensile load P_T is applied to the top block (Fig. 1), the load is supported by both the tie rods in tension and the soil mass in compression. That is, when P_C is applied, the tie rod tension decreases to $T - \triangle T$ ($\triangle T > 0$) associated with compression of the backfill, while the increase in the vertical load applied on the backfill is reduced to $P_C - \triangle T$ from the value P_C. A similar benefit can be expected also when P_T is applied. In addition, preloading also introduces prestress in the reinforcement, which contributes to maintaining the integrity of the backfill. Under high confining pressure by prestressing, a high shear rigidity of the backfill is also attained.

Full-scale models and the first prototype

Three full-scale models and one prototype bridge pier, as listed in Table 1, have been constructed.

Table 1. Three full-scale models and the prototype pier

	3M	3A	3B	Pier
γ_d [1]	18.4 kN/m³ (1.88 tf/m³)	20.5 kN/m³ (2.09 tf/m³)		Not measured
k_{30}	160 N/cm³ (16.3 kgf/cm³) after loading tests (with gypsum)[3]	157 to 196 N/cm³ (16 to 20 kgf/cm³) (with gypsum)[3]		116 N/cm³ (11.8kgf/cm³) (with sand layer)
Material	Chiba gravel			Kyushu gravel
Geogrid v.spacing[2]	30 cm			15 cm
Backfill dimensions	3.2m x 4m x 5m(high)	2.7m x 4m x 5m(high)		6.4m x 4.4m x 2.7m(high)
Reaction block area	7.6m² = 2m x 3.8m	5.7m² = 1.5m x 3.8m		12m² =5m x 2.4m
Boundary conditions	Plane strain	Plane strain		Rectangular Prism

1) Average dry unit weight of the backfill.
2) Vertical spacing.
3) For a good contact between the plate and the ground surface.

1. Full-scale model (3M) constructed in 1995

The test section was a part of a full-scale model test facility constructed in 1995, consisting of four 5 m-high, 7.4 m-long, and 4 m-wide sections, each separated by RC walls. Vinyl sheets smeared with grease were used to decrease the friction between each RC wall and the backfill. Segment 3M was the central part of one of the sections (Fig. 2a). The backfill was a well-graded crushed sandstone (D_{max} = 37.5 mm, D_{50} = 4 mm, Uc = 6.4 mm / 0.09 mm = 71 and a fines content equal to 8.0 % with w = 7 %, see Fig. 3) was compacted to a relatively low dry unit weight of 18.4 kN/m^3 (1.88 tf/m^3). The reinforcement was a geogrid made of polyvinyl alcohol (Vinylon), having a nominal tensile rupture strength of 73.5 kN/m (7.5 tonf/m), which is widely used to construct prototype GRS retaining walls. The vertical spacing of the geogrid layers was 30 cm.

The adjacent Segments, 3S and 3N, had wrapped-around wall faces, constructed with help of gravel-filled gabions stacked on the shoulder of each soil layer. Full-height rigid facings were not constructed. A pair of 0.2 m-wide unreinforced gravel zones separate segment 3M from 3S and 3N. Moreover, before the test on 3M, vertical cracks had developed in the unreinforced zones to a depth of more than 2.5 m from the crest of the backfill by pre-loading tests on 3S and 3N. Therefore, Segment 3M can be considered to be mechanically nearly independent from 3S and 3N. The cross-sectional area of Segment 3M was 12.8 m^2(=3.2 m x 4 m); the cross-sectional area of the reaction blocks was 7.6 m^2 (= 2 m x 3.8 m).

During dismantling the test section after a series of loading tests described below, a plate loading test was performed at the center of the 4th layer (1.2 m-high) of Segment 3M. A cirlular steel plate with a diameter of 30 cm and a thickness of 2 cm was placed on the flat surface; gypsum was used between the plate and the soil surface to ensure good bedding (Fig. 4). The coefficient of vertical sub-grade reaction, k_{30}, was defined as P/(S·A) at S = 1.25 mm; where S is the plate settlement, A is the plate area (706.5 cm^2), and P is the applied load (Fig. 4). This index is widely used to control the backfill compaction in Japan. The measured k_{30} value was 160 MN/m^3 (16.3 kgf/cm^3), which is large enough for the backfill to support railway. However, this value was obtained after a number of loading tests, a much lower value would have been obtained if measured before the loading tests.

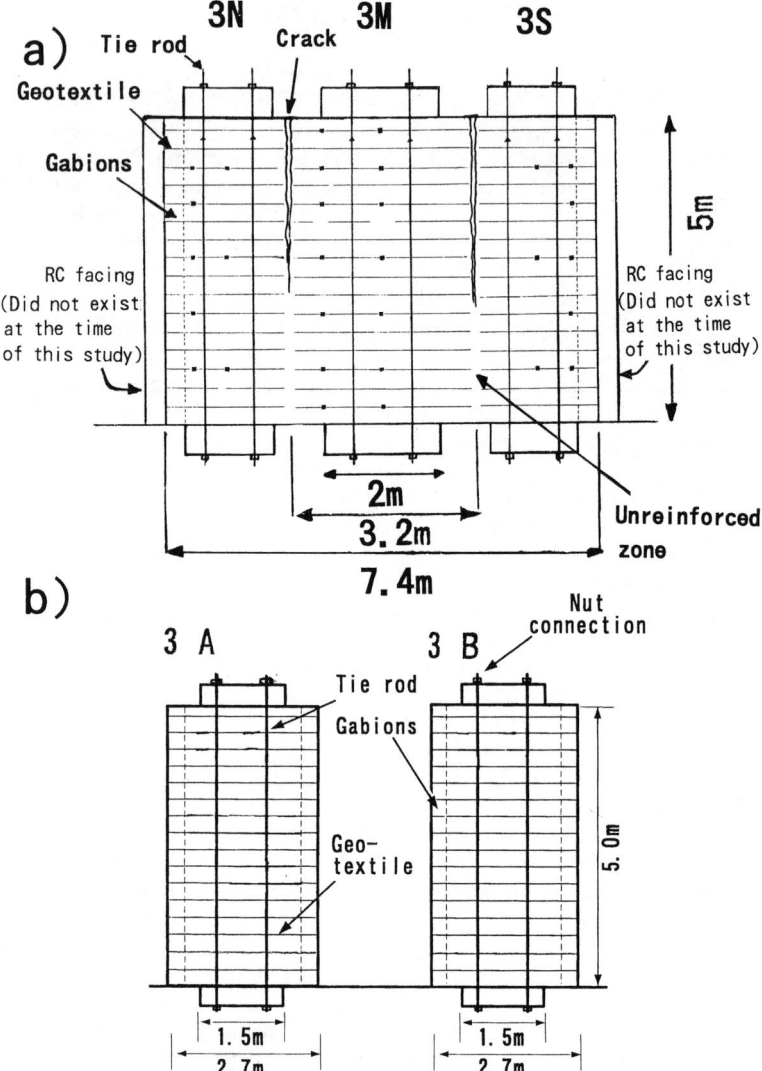

Fig.2 a) Cross-section of the test segment 3M constructed in 1995; b) cross-section of the model 3A and 3B, constructed in 1997.

2. Full-scale models (3A,3B) constructed in 1997

In 1997, two other full-scale models, 3A and 3B, were constructed in the test section where the model 3M had been constructed and dismantled (Fig. 2b). The same backfill material as 3M was compacted to a dry unit weight of 20.5 kN/m^3 (20.9 tf/m^3). The main material and mechanical properties of the geogrid were also the same as those used for 3M except the coating material. The vertical spacing of the geogrid layers was also 30 cm. Vinyl sheets smeared with grease were also placed between each RC wall and the backfill to decrease side wall friction.

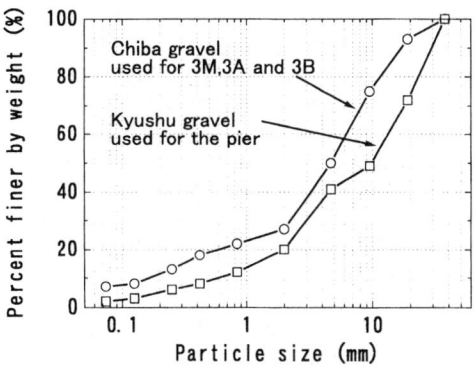

Fig. 3 Particle size of backfill materials for the model and the prototype pier.

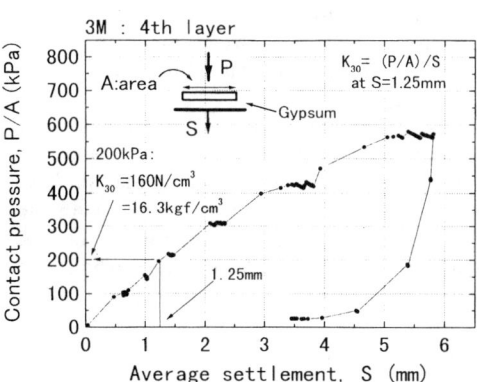

Fig. 4 Result of a plate loading test and definition of k_{30} for Segment 3M.

3A and 3B are 5 m-high, 2.7 m-long, and 4 m-wide; the cross-sectional area of the reaction blocks is 5.7 m^2 (= 1.5 m × 3.8 m). The wall faces were constructed with help of gravel-filled gabions, wrapped around with the geogrid. Full-height rigid facings have not been constructed so far in this study.

Plate loading tests were performed at several heights in 3A during construction by the same procedure as that for 3M. Very high values, k_{30} = 157 to 196 N/cm^3 (16 to 20 kgf/cm^3), were obtained. Nearly the same value of k_{30} would have been obtained if plate loading tests were performed for 3B, as the backfill material and construction procedure were the same.

3. PLPS reinforced soil railway bridge pier

As the first prototype PLPS geogrid-reinforced soil structure, a bridge pier was constructed in 1996 to support two 16.5 m-long steel bridge girders for a single railway track in Fukuoka City, Japan (pier P1 in Figs. 5a) (Uchimura et al., 1998). The bridge has been opened to service since August 1997, and will be used for three years. The cross-section of the pier is 6.4 m x 4.4 m, and the height of the backfill is 2.7 m; the cross-sectional area of the top reaction block is 12 m^2 (= 5 m x 2.4 m). The design dead load by the girder weight and design live load by train loads including impact load are 196 kN and 1,280 kN, respectively.

A well-graded gravel of crushed sandstone (D_{max} = 30 mm, D_{50} = 0.9 mm, U_c = 16.5, see Fig. 3) was used. A hand-operated 30 kg-vibration compactor and a hand-operated 60 kg-tamper were used for compacting the backfill; however, the unit weight of the compacted backfill was not measured. The geogrid is the same as that used in the model 3M; but the average vertical spacing is 15 cm. The wrapped-around wall faces were constructed with help of gravel-filled gabions on each layer around the pier.

Plate loading tests were performed at the center of the backfill crest. An about 1 cm-thick layer of poorly graded fine sand was used for a good bedding between the plate and ground surface, instead of using gypsum. This procedure might have made the value of k_{30} smaller due to a lower stiffness of the sand layer. A value of k_{30} = 116 N/cm^3 (11.8 kgf/cm^3) was obtained, which is a typical value for actual railway foundations.

Loading conditions for the models and the pier

Figs. 6 to 9 shows the time histories of the load applied on the top of the backfill and the vertical compression of the backfill and the relationships between them for the four cases described above. The applied load is equal to either 1) tie rod tension applied by using hydraulic jacks set at the top of the tie rods for models 3A, 3B and the pier; or 2) the sum of the tie rod tension (subjective) and the compressive load applied by using hydraulic jacks on the reaction block for the case of model 3M.

The backfill of 3M was first preloaded for 4 days; first some creep deformation was allowed for 2 hours at 730 kN (74tf) (Stage 2 in Fig. 6), and 980 kN (100tf) was applied for 8 minutes (3). Then after unloaded to 640 kN (65 tf), stress relaxation was observed for 24 hours (4). Next, load of 980 kN was applied again for 72 hours (5), which was unloaded to 640 kN, and then the backfill was left prestressed for 14 months (6). During this period, the prestress in the backfill decreased by 20 % due to soil relaxation. After that, one cycle of loading and unloading was applied on the top reaction block with the tie rod tension (prestress) working, followed by constant load for one day, and one hundred cycles of loading applied with an amplitude of by 196

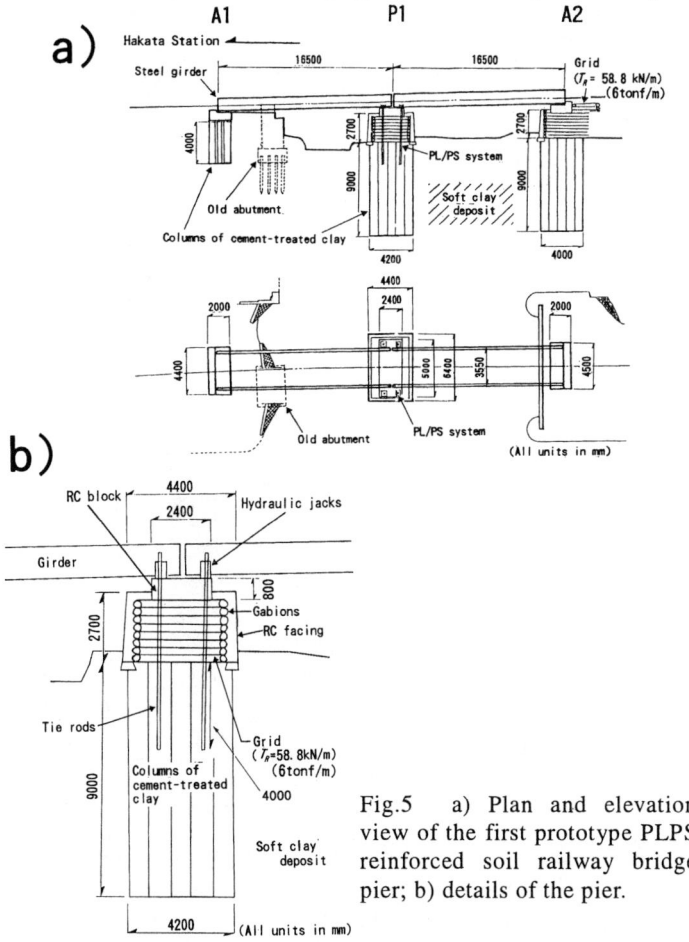

Fig.5 a) Plan and elevation view of the first prototype PLPS reinforced soil railway bridge pier; b) details of the pier.

kN (20 tf), 392 kN (40 tf), and 588 kN (60 tf) respectively (7 to 9 in Fig. 6). Then a sequence of loading and unloading with constant peak load of 1180kN (120tf) was applied (10) for one day. Because of these external loads, the tie rod tension decreased quickly, corresponding to the associated backfill compression.

On the 740th day, five series of cyclic loading were applied again on the top reaction block (11 to 15); each series consisted of 50 cycles of loading and unloading with an amplitude of 392 kN or 1180 kN at each load level.

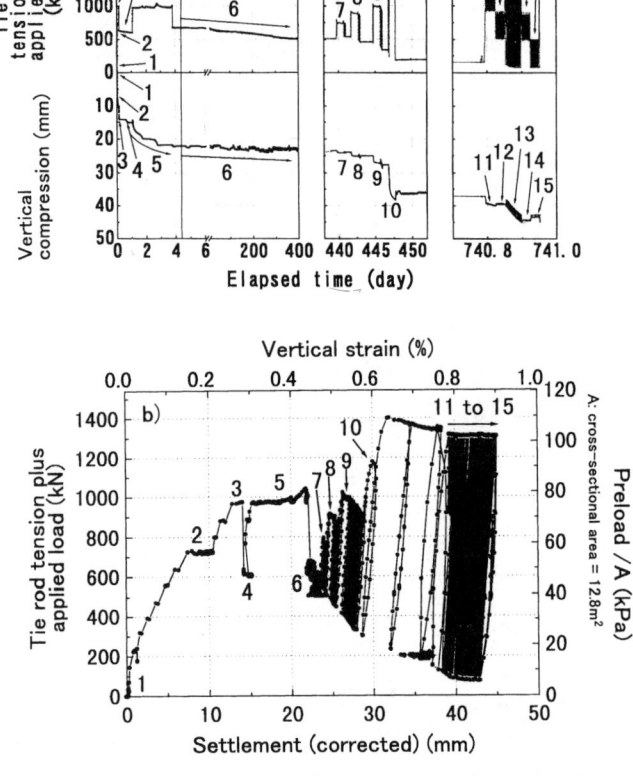

Fig. 6 Test results of the model 3M: a) time histories load applied on the top of the backfill and vertical compression; and b) relationships between the applied load and the compression.

The backfill of 3A was preloaded stepwise; each step consisted of one hour of constant load allowing creep deformation and 3 cycles of cyclic loads with a small double amplitude of 196 kN, applied to investigate the elastic stiffness. Constant load of 1180 kN (120 tf) was applied first for four days (1 in Fig. 7), and then five days (2) and two days (3) with a sequence of full unloading and reloading between them. Then, 2350 kN (240 tf) was applied for six days (4) and one day (5). Finally, 50 cyclic loads with a large double amplitude of 2350 kN were applied (6).

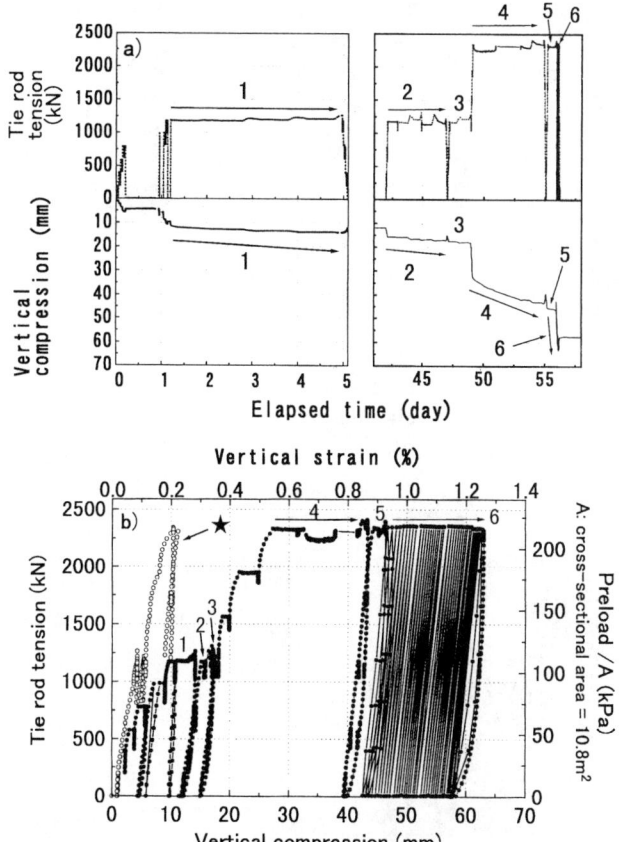

Fig. 7 Test results of the model 3A: a) time histories of load applied on the top of the backfill and vertical compression; and b) relationships between the applied load and the compression.

The backfill of 3B was preloaded with cyclic loads at various intermediate load levels(Fig. 8); each cyclic load consists of 50 cycles of a double amplitude of 196 kN (20 tf), 392 kN (40 tf), or 2350 kN (240 tf).

The backfill of the railway bridge pier was preloaded up to 2350 kN (240 tf). Full-height rigid facings were cast-in-place after the preloading stage. In the first day, the backfill was preloaded stepwise up to 1,960 kN (up to Stage 1 in Fig. 9); each step consists of a load increment of 196 kN (20 tf), applied within 2

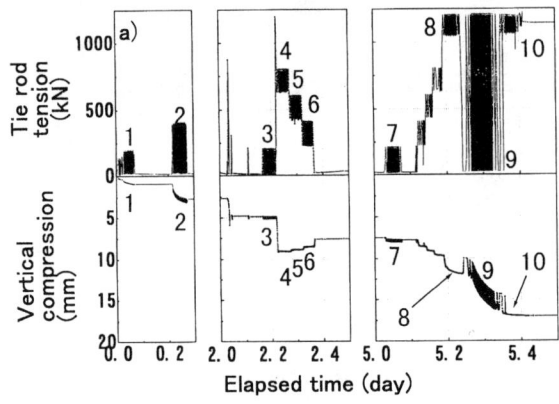

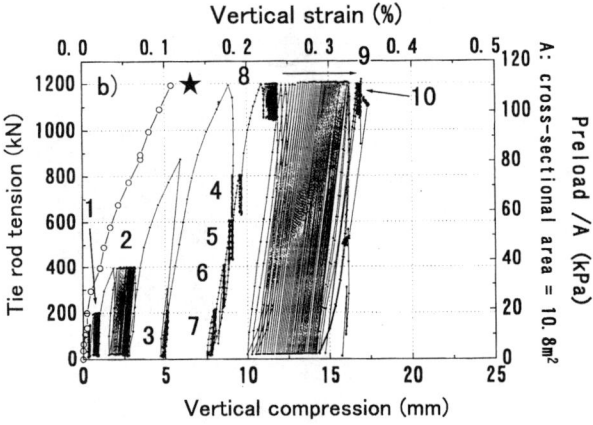

Fig. 8 Test results of the model 3B: a) time histories of load applied on the top of the backfill and vertical compression; and b) relationships between the applied load and the compression.

minutes or less, followed by a pause keeping the load constant for 30 or 60 minutes. In the fifth day, the load was decreased to 905 kN (2), followed by reloading. In the sixth day, the load was increased to 2,350 kN (3). In the seventh day, the load was decreased to zero (4), followed by full reloading. In the tenth day, the load was decreased to about 1,100 kN (5) and then the backfill was left under prestressed conditions for three days. Finally, the load was increased to 2,350 kN (6), maintained for three hours, decreased again to 980 kN (7), and then, the backfill was brought to the final prestressed condition. On the 20th day, the RC fac-

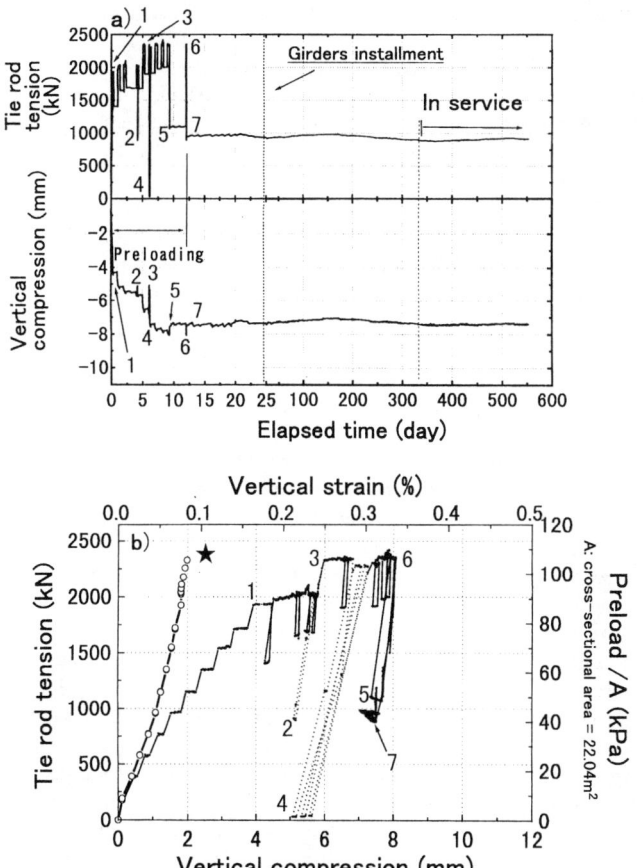

Fig. 9 Behavior of the prototype pier: a) time histories of load applied on the top of the backfill and vertical compression; and b) relationships between the applied load and the compression.

ings were constructed; and on the 25th day, the girders were placed, whose load on the pier was 210 kN (21.5tf). After that, the pier was left without live load for more than nine months. Then, trains started to run over more than 125 times everyday; the train load on the pier is about 392 kN (40 tf). Before and after opening to service, the backfill did not show any noticeable creep and residual compression while the prestress remained nearly constant. The transient compression when a train passes over is only 0.02 mm (Uchimura et al., 1998).

Comparison of the behaviors of the models and the pier

Fig. 10 compares the relationships between the average stress applied as preload, and the average vertical strain of the backfill of the three models and the pier; the unloading and reloading parts have been deleted to make the figure legible. The stress and strain relationships for 3A, 3B, and the pier are nearly the same, while the stiffness of the backfill of 3M is nearly half of the others. One of the most possible reasons for this difference is a low unit weight of the backfill of 3M (see Table 1). The dry unit weight of the backfill of 3A and 3B are nearly the same and much higher. Although the backfill unit weight of the pier was not measured, the backfill is considered to be very dense when based on the measured k_{30} value. This comparison suggests the importance of good backfill compaction.

The curve denoted by the symbol ★ in Fig. 7b is the relationship between the average stress and the integrated value of the instantaneous compressive strain increments that occurred during the primary loading process. The difference between the total compressive strains and the instantaneous compressive strains consists of: 1) viscous compression caused by creep deformation; and 2) residual compression caused by cyclic loading. The difference increases with the load level, and 85 % of the largest total compression is the irreversible component. Similarly, the integrated compression, obtained by excluding the residual component caused by cyclic load, is denoted by ★ in Fig. 8. The residual compression by cyclic load is more than 60 % of the largest total compression. The curve denoted by ★ in Fig. 9 shows also the integrated value of the instantaneous compression increments occurred at each loading step. More than 70 % of the largest total compression is due to creep deformation. It is to be noted that the backfills of 3A, 3B, and the pier showed very large irreversible compression caused by creep deformation and residual deformation by cyclic loading, although they had been compacted very on loading conditions. Part or most of the creep and residual deformations observed would have occurred as plastic ones if loading were monotonic without creep stages and cyclic loading.

It is seen from Figs. 6 and 8 that, with respect to the amount of residual strain caused by cyclic loads:
1) in case the backfill has been loaded to the highest level in its loading history and not unloaded, even cyclic loading with a small amplitude (196 kN) can cause large residual compression (11 in Fig. 6; 1,2, 8 and 10 in Fig. 8);
2) in case the backfill has been unloaded to a certain prestress level (not to zero), the residual compression caused by cyclic loading can become much smaller, and the amplitude of the transient compression can also become very small (12 and 14 in Fig. 6; 4,5, and 6 in Fig. 8);
3) in case the backfill has been fully unloaded to zero, the amplitude of the compression by cyclic load become much larger, while the residual compression remains small (15 in Fig. 6; 3, and 7 in Fig. 8); and
4) in case the amplitude of the cyclic load is very large close to the preload level, both the residual and transient compression can be very large (13 in Fig. 6; 9 in Fig. 8).

These results suggest that the preload should be sufficiently larger than the assumed load level in service. And the preload should not be unloaded to zero; some prestress should remain. However, the prestress should not be very high so that the maximum stress during cyclic loadings does not approach the maximum stress during preloading. At highest, the prestress should be less than the preload level by the assumed loads in service. Cyclic preloading may be effective to compresses the backfill in a short construction period.

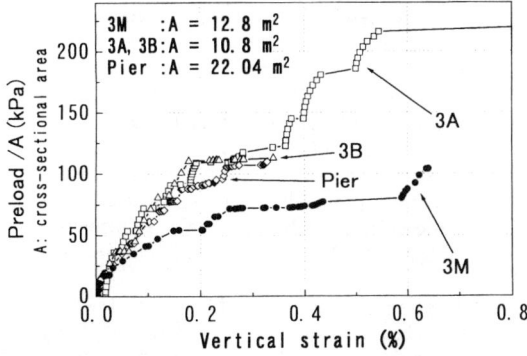

Fig. 10 Comparison of relationships between average stress and average strain of the models and the prototype pier.

Conclusions

The deformation properties of the first prototype PLPS reinforced soil structure and three full-scale models of PLPS embankment were compared. It was confirmed that the compaction is essential to make the backfill stiff. It was found, however, that even a very well compacted backfill could exhibit very large irreversible deformation by static (i.e. monotonic and creep) and transient cyclic loading. To restrain this irreversible deformation, it is effective to preload the backfill sufficiently and to unload some part of it, and to keep high prestress in the backfill.

Acknowledgement

The authors are deeply indebted to their colleagues in helping us with this study, particularly, Dr. J.Koseki, Dr. T.Kodaka, Mr. T.Sato, in Institute of Industrial Science, University of Tokyo, and Mr. K.Kojima and Mr. H.Kimura in Railway Technical Research Institute. Mr. T.Maeda and Mr. H.Tsuru in Kyushu Railway Company helped us as the project managers for the PLPS bridge pier.

References

Tatsuoka,F. and Kohata,Y. (1995): Stiffness of hard soils and soft rocks in engineering applications, Proc. Int. Symp. on Pre-Failure Deformation Characteristics of Geomaterials, IS Hokkaido '94, Balkema, Vol.2, pp.947-1063

Tatsuoka,F., Uchimura,T., Tateyama,M. and Muramoto,K. (1996): Creep Deformation and Stress Relaxation in Preloaded/Prestressed Geosynthetic-Reinforced Soil Retaining Walls, Session "Measuring and Modeling Time-Dependent Soil Behavior", ASCE Washington Convention, Geotechnical Special Publication, No. 61, pp.258-272.

Tatsuoka, F., Uchimura,T., and Tateyama, M.(1997a): Preloaded and prestressed reinforced soil, Soils and Foundations, Vol.37, No.3, pp.79-94

Tatsuoka,F., Tateyama,M., Uchimura,T. and Koseki,J. (1997b): Geosynthetic-Reinforced Soil Retaining Walls as Important Permanent Structures, The 1996-1997 Mercer Lecture, Geosynthetics International, Vol.4, No.2, pp.81-136.

Uchimura,T., Tatsuoka,F., Sato,T., Tateyama,M. and Tamura,Y. (1996). Performance of preloaded and prestressed geosynthetic-reinforced soil, Proceedings of International Symposium on Earth Reinforcement, Fukuoka, Balkema (Ochiai et al., eds), Vol. 1, pp.537-542.

Uchimura,T., Tatsuoka,F., Tateyama,M., Koga,T. (1998): Preloaded-Prestressed Geogrid-reinforced Soil Bridge Pier, Proceedings of the 6th International Conference on Geosynthetics, Atlanta, Vol.2, pp.565-572.

Slip-Line Analyses of Geosynthetic-Reinforced Strip Footings

Aigen Zhao[1], Member, ASCE

Abstract

Theoretical analyses of the plastic failure region and the ultimate bearing capacity of reinforced soils under strip footings are presented in this paper. The analysis is based on the failure criteria for homogenized reinforced soils and the application of the slip-line method (the method of characteristics). Two ideal footing bases are considered: a perfectly smooth base and a perfectly rough base. Influence of geosynthetic reinforcement properties (tensile strength, deployment orientation) and soil strength (friction angle and cohesion) on the bearing capacity of reinforced footings and the plastic failure region are investigated. Three design examples of strip footings are analyzed: an unreinforced footing with significantly deep excavation; a geosynthetic reinforced footing back filled with the same subgrade soil, and a geosynthetic reinforced footing back filled with better quality soil. The bearing capacity and stress characteristics field are presented. This analysis can lead to a better understanding of the plastic failure region of reinforced footings and more rigorous design methods.

Introduction

Geosynthetic-reinforced footings have been put into practice in recent years. Some experimental tests on geosynthetic reinforced foundations were reported by Guido et. al (1987), Khing et. al (1993), Omar et. al (1993), and Adams and Collins (1997) though mainly focused on quantifying the contribution of geosynthetic reinforcement to the bearing capacity. Test results and a design procedure for layered soil with a single layer of geosynthetic reinforcement were presented by Ismail and Raymond (1995). A design method proposed by Zhao et. al (1996) was based on the slip-line method and the assumption of a smooth footing base, where reinforcement strength, length, spacing, and the number of reinforcement layers were discussed. Research of the plastic failure region and the bearing capacity of a reinforced footing under a rough base assumption has not been explored.

---

[1] Technical Director, Tenax Corporation, 4800 East Monument Street, Baltimore, MD 21025

In this paper, The slip-line method is first applied to analyze unreinforced footings, and then the results are compared to the solution from the traditional bearing capacity formula to verify the algorithm. Two ideal footing bases are considered: a perfectly smooth base and a perfectly rough base. The slip-line method is then used to analyze geosynthetic-reinforced footings to investigate the influence of geosynthetic properties (tensile strength, orientation) and soil properties (internal friction angle, and cohesion) on the bearing capacity and the plastic failure region underneath reinforced footings. Finally, three design examples for bearing capacity of strip footings are presented, an unreinforced footing with significantly deep excavation; a geosynthetic reinforced footing back filled with the same subgrade soil, and a reinforced footing back filled with better quality soil. Both bearing capacity and plastic failure region are discussed for these three cases. The paper is finished with concluding remarks.

Slip-Line Equations for Reinforced Soils

The plastic stress state of the soil is assumed to conform to the Mohr-Coulomb failure condition, and Tresca failure condition is assumed for the reinforcement. Under plane strain conditions, the failure criterion for a homogenized reinforced soil composite was derived and presented by Michalowski and Zhao (1995). The failure criteria are anisotropic (dependent on the major principal stress direction). The failure criteria for reinforced soils along with stress equilibrium equations lead to a set of hyperbolic-type partial differential equations, which can be solved using the slip-line method. The slip-line method is considered as a rigorous approach to analyze plastic region of soil structures since it satisfies the local stress equilibrium. Following Booker and Davis (1972), the equations of characteristics for anisotropic soils can be expressed as

$$\frac{dy}{dx} = \tan(\psi - m - v) \quad \text{(characteristic } s_1)$$

$$\frac{dy}{dx} = \tan(\psi - m + v) \quad \text{(characteristic } s_2) \tag{1}$$

And the stress relations along characteristic s_1 and s_2 are

$$\sin[2(m-v)]\frac{\partial p}{\partial s_1} + 2F(p,\psi)\frac{\partial \psi}{\partial s_1} + \gamma \cos(2m)[\cos(2v)\frac{\partial x}{\partial s_1} - \sin(2v)\frac{\partial y}{\partial s_1}] = 0$$

$$\sin[2(m+v)]\frac{\partial p}{\partial s_2} + 2F(p,\psi)\frac{\partial \psi}{\partial s_2} + \gamma \cos(2m)[\cos(2v)\frac{\partial x}{\partial s_2} + \sin(2v)\frac{\partial y}{\partial s_2}] = 0 \tag{2}$$

Where $F(p, \psi)$ is the failure function for reinforced soils (reinforcement is uni-directionally oriented and evenly spaced), p is the stress parameter equal to the mean of the maximum and minimum principal stresses. ψ is the angle of major principal stress with respect to the x-axis, γ is the unit weight of the soil, m and υ are described by

$$\tan(2m) = \frac{1}{2F(p,\psi)} \frac{\partial F(p,\psi)}{\partial \psi}$$

$$\cos(2v) = \cos(2m)\frac{\partial F(p,\psi)}{\partial p}$$

(3)

For isotropic soils (m = 0 and $\upsilon = \pi/4 - \phi/2$, where ϕ is the internal friction angle of the soil), it can be shown that the above stress characteristics and stress relations are reduced to classic slip-line equations by Sokolovskii (1965).

Slip-Line Analyses of Strip Footings

To verify the algorithm of the proposed slip-line method, the bearing capacity and plastic failure region of an unreinforced footing are first investigated, and then compared to the result from a traditional bearing capacity formula. For the purpose of demonstration, an example problem (see Figure 1) is calculated, where the footing width B = 2 m, footing excavation depth D =0.5m, soil internal friction angle $\phi = 25°$, soil cohesion c = 0, soil unit weight $\gamma = 18$ kN/m^3. Two types of ideal footing bases: a perfectly smooth base footing and a perfectly rough base footing, are considered in the slip-line solution.

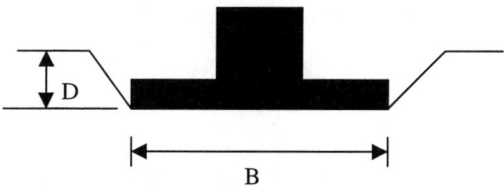

Figure 1. Slip line analysis of a strip footing

Smooth base assumption significantly simplifies numerical calculations since the major principal stress direction is vertical ($\psi = \pi/2$) along the base AB (see Figure 2). Calculations starts with AC boundary, where the direction of the major principal stress with respect to x-axis $\psi = 0$, the stress parameter p = γ D/(1-sinϕ). Having determined the stress boundary condition at AC, the Cauchy boundary value problem was solved first in the region ACD, followed by the characteristics problem in area ADE with a singular point at A, and then the problem with mixed boundary conditions in ABE. Once p along the base AB is found, the bearing capacity of the footing can be calculated. The calculated bearing pressure on the footing base is non-uniformly distributed, as indicated in Figure 2. The average bearing capacity of the footing from the slip-line method is $q/\gamma B = 5.11$.

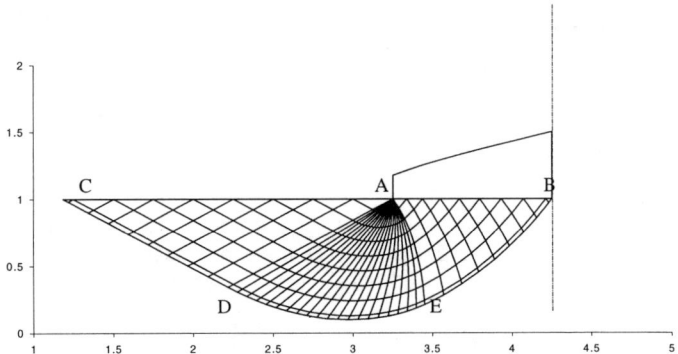

Figure 2. Stress characteristics field for an unreinforced footing (smooth base) (D/B =0.25, $\phi=25°$, $q/\gamma B$ =5.11)

Under a perfectly rough base assumption, calculations for the characteristics boundary value problem and the mixed boundary value problem are more complicated since neither the major principal stress direction nor the mean stress parameter p is known along the base. The last stress characteristic s_2 in the characteristic boundary value problem must intersect the symmetry line at an angle of $(\upsilon\text{-m})$ ($\pi/4 - \phi/2$ for unreinforced soils). The average bearing capacity of the footing in this case is $q/\gamma B$ =7.59. The stress characteristics field is shown in Figure 3.

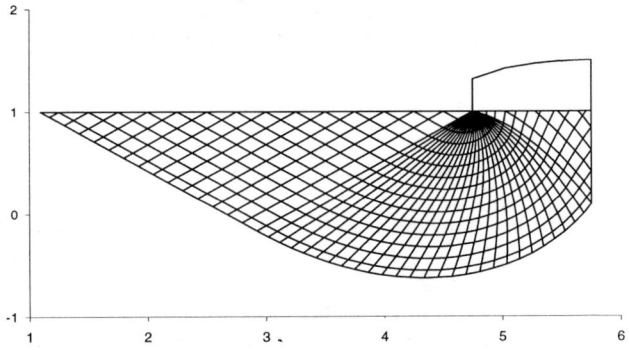

Figure 3. Stress characteristics field for an unreinforced footing (rough base) (D/B =0.25, $\phi=25°$, $q/\gamma B$ =7.6)

For unreinforced strip footings, the bearing capacity can also be calculated by classic bearing capacity equations. For granular soils,

$$q = \gamma D N_q + \frac{1}{2}\gamma B N_\gamma \tag{4}$$

Where D=0.5m, B=2m, γ=18kN/m^3; N_q and N_γ are bearing capacity factors, which are functions of the internal friction angle of the foundation soil, according to Hanson (1970)

$$N_q = e^{\pi \tan\phi} \tan^2(45^o + \frac{\phi}{2})$$

$$N_\gamma = 1.5(N_q - 1)\tan\phi \tag{5}$$

In this example, $\phi = 25°$, therefore,

$$N_q = e^{\pi \tan(25^o)} \tan^2(45^o + \frac{25^o}{2}) = 10.64$$
$$N_\gamma = 1.5(N_q - 1)\tan 25^o = 6.74$$

Thus

$$\begin{aligned} q &= \gamma D N_q + \frac{1}{2}\gamma B N_\gamma \\ &= 18(0.5)(10.64) + 0.5(18)(2)(6.74) \\ &= 217\ kPa \end{aligned}$$

q/γB = 217/36=6.02

The above result from the bearing capacity equation falls into the ranges calculated by the slip-line method (q/γB = 5.11 under a perfectly smooth base assumption and q/γB = 7.6 under a perfectly rough base assumption). The bearing capacity and the size of the plastic failure region from a smooth base assumption are significantly smaller than that calculated from the rough base assumption. Rough base assumption is believed to be more realistic compared to a smooth footing assumption.

Figures 4 and 5 present slip-line solutions for a geosynthetic--reinforced footing under a perfectly smooth base assumption and a perfectly rough base assumption respectively. The long term design strength of geosynthetic reinforcement is T=30 kN/m. The reinforcement is horizontally placed underneath the footing at an even spacing of s=0.5m. The average bearing capacity of the footing with smooth base is q/γB = 11.41, while the bearing capacity of the footing with a rough base is q/γB = 14.6 (a 28% increase over a smooth base condition). The size of the plastic failure region under the footing with a rough base is also significantly larger than that with a smooth base (70% larger in horizontal direction and almost 100% in the vertical direction). Similar to unreinforced footings, rough base assumption is considered as a more realistic modeling compared to smooth base assumption. All the following calculations will be based on a rough footing assumption.

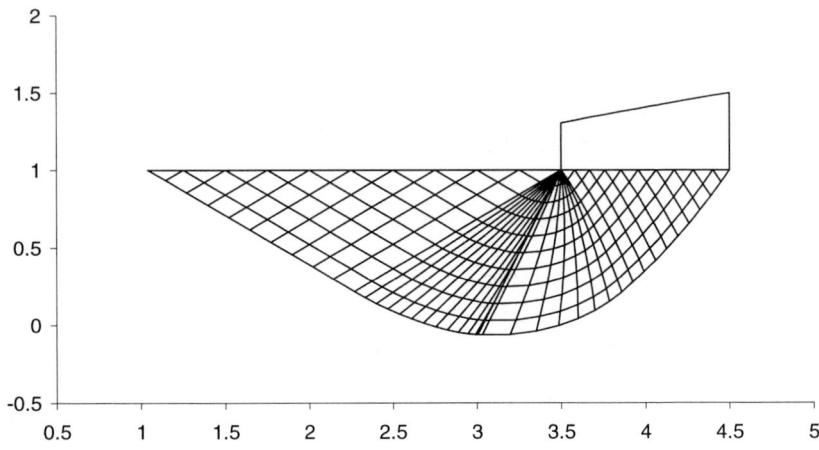

Figure 4. Stress characteristics field for a reinforced footing (smooth base)
(D/B=0.25, $\phi=25°$, T=30 kN/m, s=0.5m, $q/\gamma B$ =11.41)

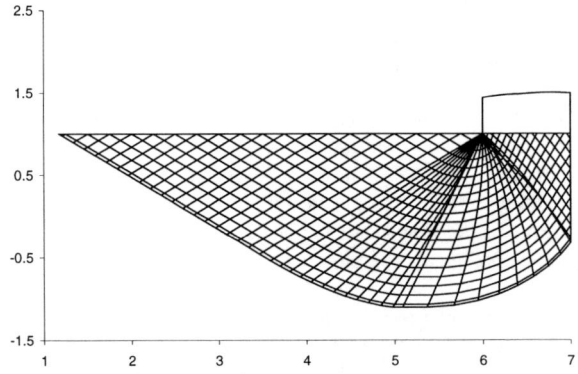

Figure 5. Stress characteristics field for a reinforced footing (rough base)
(D/B=0.25, $\phi=25°$, T=30 kN/m, s=0.5m, $q/\gamma B$=14.59)

Figure 6 shows the stress characteristics field for a footing with soil friction angle of 35°. Similar to the bearing capacity of unreinforced footings, the effect of soil friction angle on the bearing capacity of geosynthetic-reinforced footings is also significant. When the friction angle is increased from 25° to 35°, bearing capacity is increased by three times. The size of the plastic failure region is also increased by 44% in the horizontal direction and 22% in the vertical direction, respectively.

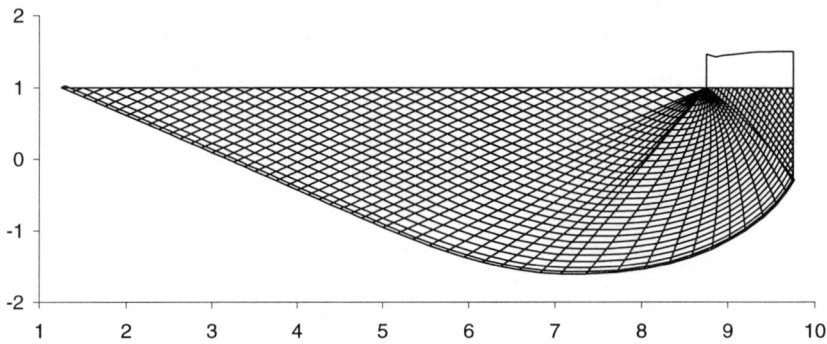

Figure 6. Stress characteristics field for a reinforced footing (rough base)
(D/B=0.25, ϕ=35°, T=30 kN/m, s=0.5m, q/γB =45.83)

The slip-line field of a foundation soil with cohesion c=75 kPa is presented in Figure 7. The bearing capacity is increased 4 times over the granular soil case as shown in Figure 5. The size of the plastic failure region is also increased by 19% in the horizontal direction and 18% in the vertical direction. The influence of soil cohesion on the bearing capacity and the plastic failure region of a reinforced footing is significant.

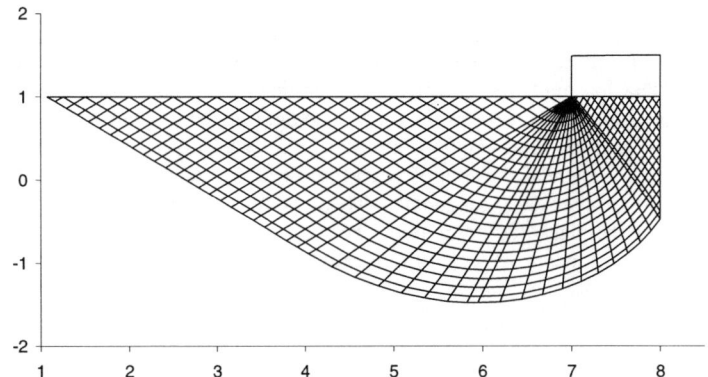

Figure 7. Stress characteristics field for a reinforced footing (rough base)
(D/B=0.25, ϕ=25°, c=75 kPa, T=30 kN/m, s=0.5m, q/γB =58.47)

Figure 8 shows the stress characteristics field for a strip footing reinforced with geosynthetic placed at an inclination angle of $\alpha=10°$ to the horizontal direction. The bearing capacity is reduced by 10% compared to the case in Figure 5 where reinforcement is horizontally placed. The size of the plastic failure region is also deceased by 5% in both horizontal and vertical directions. It indicates that horizontal direction is the optimal direction to place geosynthetic reinforcement underneath a strip footing. It is obvious that geosynthetic reinforcement as a tensile element is most effective when they are placed in the maximum tensile strain direction.

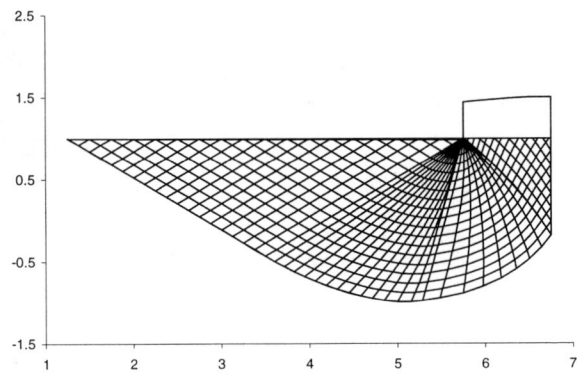

Figure 8. Stress characteristics field for a reinforced footing (rough base)
(D/B=0.25, $\phi=25°$, $\alpha=10°$, T=30 kN/m, s=0.5m, $q/\gamma B$ =13.13)

Figure 9 presents the stress characteristics field for a reinforced strip footing with reinforcement of higher tensile strength. The reinforcement strength is doubled compared to the case in Figure 5. The tensile strength of reinforcement is now T=60 kN/m. The spacing of reinforcement is kept the same. The bearing capacity increase contributed by the increase in the tensile strength of reinforcement is 44%, while the size of the plastic failure region is almost unchanged. This indicates that by increasing the tensile strength of the reinforcement, the bearing capacity is not increased at the proportion of the same magnitude. There is little influence of the reinforcement strength on the plastic failure region. Therefore, the design of the reinforcement length and placement depth is not significantly affected by the tensile strength of the reinforcement.

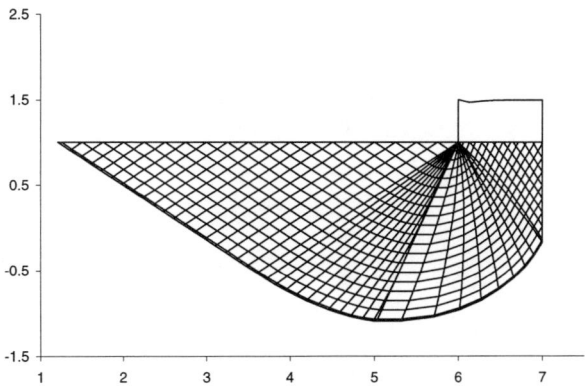

Figure 9. Stress characteristics field for a reinforced footing (rough base)
(D/B=0.25, $\phi=25°$, T=60 kN/m, s=0.5m, $q/\gamma B=21$)

Design Examples

Three design examples for strip footings are presented. One is for an unreinforced footing design, and the other two are for reinforced footing design. The design focus is to determine the required depth of excavation with and without geosynthetic reinforcement. The same data from the previous example problem as shown in Figure 1 are used. The vertical load on the footing is assumed to be 350 kPa, the target safety factor for bearing capacity Fs = 2, therefore, the design bearing capacity of the footing is 2*350 kPa = 700 kPa. The bearing capacity of the foundation soils calculated in the previous section is 217 kPa (D=0.5m). In order to increase the bearing capacity from 217 kPa to 700 kPa, at least two possible approaches are feasible without reinforcement: either to increase the depth of excavation D or increase the footing width. The excavation depth necessary to achieve the design bearing capacity is described here. Solutions from both classic bearing capacity equation and the slip-line method are presented.

The excavation depth required to meet the bearing capacity can be calculated by Henson's bearing capacity formula as follows:

$700 = 18 \cdot D \cdot 10.64 + 0.5 \cdot 18 \cdot 2 \cdot 6.74$

$$D = \frac{700 - 121.32}{191.52} = 3m$$

The slip-line solution (see Figure 10) shows the stress characteristics field under an excavation depth D=2.5m. The bearing capacity is 702 kPa. It indicates that to achieve the required bearing capacity, the excavation depth has to be increased from D=0.5m to D=2.5m.

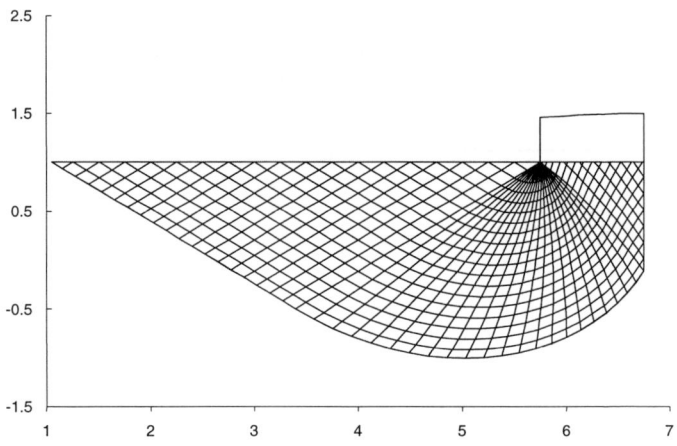

Figure 10. Stress characteristics field for a reinforced footing (rough base) (D = 2.5m, ϕ = 25°, q=702 kPa)

An alternative footing design is to use geosynthetic reinforcement to increase the bearing capacity while keeping the excavation depth unchanged (D=0.5m). The analysis considered here is to use the existing foundation soil as the back fill material. The soil has a friction angle of 25° and a unit weight of 18 kN/m³. Geosynthetic reinforcement has a long term design tensile strength of T = 33 kN/m, and the reinforcement is horizontally placed at an even spacing of s = 0.3m. The bearing capacity from the slip-line method is 716 kPa, which meets the target bearing capacity. The slip-line field is shown in Figure 11.

The third design example is to use higher friction fill soil to reduce the reinforcement strength and/or spacing requirements. In this case, back fill soil is assumed to have a friction angle of 30° (a 20% increase over the case in Figure 11). The long-term design strength of reinforcement is now reduced to T=15 kN/m and a spacing is increased to s=0.5m. The bearing capacity from the slip-line method is 727 kPa, which meets the bearing capacity requirement. The stress characteristics field is presented in Figure 12.

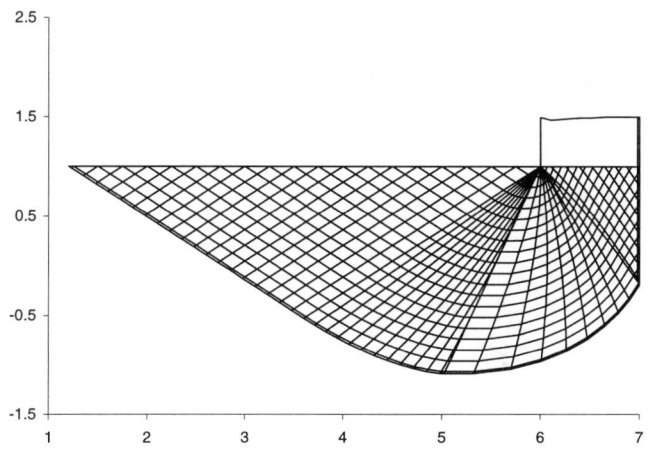

Figure 11. Stress characteristics field for a reinforced footing (rough base)
($D = 0.5$m, $\phi = 25°$, T=33 kN/m, s=0.3m, q =716kPa)

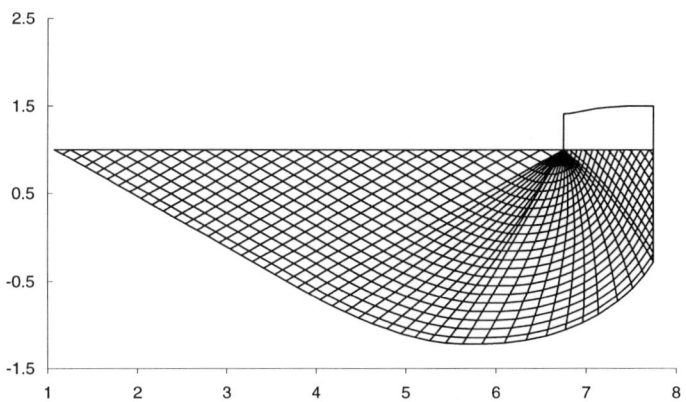

Figure 12. Stress characteristics fields for a reinforced footing (rough base)
($D = 0.5$m, $\phi = 30°$, T = 15 kN/m, s = 0.5 m, q =727 kPa)

Concluding Remarks

The theoretical analyses of reinforced strip footings is based on the failure criteria for homogenized reinforced soils and the application of the slip-line method. Two footing base conditions were investigated; a perfectly smooth base and a perfectly rough base. Rough footing base is recommended for slip-line analysis for reinforced strip footings. The influence of soil friction and soil cohesion on the bearing capacity and plastic failure region of reinforced footings are found to be significant. By increasing the tensile strength of geosynthetic reinforcement, the bearing capacity of strip footing is increased but not at the same proportion of magnitude. The size of the plastic failure region under a reinforced footing is almost not affected by the tensile strength of the reinforcement. Reinforcement is most effective when it is placed in the horizontal direction under the footing. Three examples for strip footing design are presented, an unreinforced footing with a significantly deep excavation, a geosynthetic-reinforced footing back filled with the same foundation soils, and a reinforced footing back filled with a quality soil. Geosynthetic reinforcement can reduce the excavation depth to achieve the design bearing capacity. Quality fill material reduces reinforcement strength and/or spacing requirements.

The slip-line solution presented in this paper are based on rigid plasticity theory and a homogenization technique. Therefore, it is only applicable to strip footings reinforced with evenly distributed reinforcement of the same tensile strength. It does not take layered soils into account. Further studies are needed to experimentally verify the theoretical findings and develop design charts capable of determining reinforcement strength, length, spacing and the number of layers.

References

Booker, J.R. and Davis, E.H. (1972). A general treatment of plastic anisotropy under conditions of plane strain. J. Mech. Phys. Solids, 20, 239-250.

Guido, V.A. and Knueppel, J.D. and Sweeny, M.A. (1987). Plate loading tests on geogrid-reinforced earth slabs. Proc. Geosynthetics' 87, New Orleans, Vol. 1, 216-225.

Hansen, J.B. (1970) A revised and extended formula for bearing capacity, Danish Geotechnical institute, Bulletin No. 28, Copenhagen.

Ismail, I. and Raymond, G.P. (1995). Geosynthetic reinforcement of granular layered soils. Proc. Geosynthetics' 95, Nashville, Vol. 1, 317-330.

Khing, K.H., Das, B.M., Puri,V.K., Cook, E.E., Yen, S.C. (1993). The bearing-capacity of a strip foundation on geogrid-reinforced sand. Geotextiles and Geomembranes, 12, 351- 361.

Ismail, I. and Raymond, G.P. (1995). Geosynthetic reinforcement of granular layered soils. Proc. Geosynthetics' 95, Nashville, Vol.1, 317-330.

Adams M. T. and Collins, G. G. (1997). Large model spread footing load tests on geosynthetic reinforced soil foundations, J.Geot. Eng., ASCE, 123, 66-72.

Michalowski, R.L. and Zhao, A. (1995). Continuum versus structural approach to stability of reinforced soil structures, J.Geot. Eng., ASCE, 121, 152-162.

Omar, M.T., Das, B.M., Puri, V.K. and Yen, S.C. (1993). Ultimate bearing capacity of shallow foundations on sand with geogrid reinforcement. Can.Geotech. J., 30, 545-549. Sokolovskii, V. V. (1965).

Zhao, A., Montanelli, F., Rimoldi, P. (1996). Design of reinforced foundations by the slip-line method, Proc. Of Earth Reinforcement, Japan. Ochiai, Yasufuku & Omine (eds), Balkema, Rotterdam, 709-714.

Sokolovskii, V. V. (1965). Statics of granular media, Pergamon press, New York.

A Study of Stress Distribution in Geogrid-Reinforced Sand

M. A. Gabr[1], M. ASCE, Robert Dodson[2], and James G. Collin[3], M. ASCE

Abstract

Work conducted in this research aims at measuring the stress distribution with depth during a plate load testing program on geogrid-reinforced sand. A total of five load tests are performed; four on geogrid-reinforced sand and one on unreinforced sand. The tests are performed in a 1.52m x 1.52m x 1.37m(length x width x depth) steel box with a plate dimension of 0.33 m x 0.33 m. Results indicated that the magnitude of measured stresses for the reinforced sand was reasonably predicted using the Westergaard method for applied surface pressure of 28.7 kPa and 229.8 kPa, respectively. At the high stress of 430.5 kPa, methods based on both of Boussinesq and Westergaard distributions overestimated the data measured from the reinforced tests. Reducing the data in accordance with the approximate method (simplified load spreading), higher values of the angle of the stress distribution (α) were estimated for the reinforced sand as compared to the unreinforced samples which maybe indicative of a better attenuation of the stresses due to the inclusion of the reinforcement.

---

[1]

Associate Professor, Department of Civil Engineering, Campus Box 7908, North Carolina State University, Raleigh, NC 27695-7908

[2]

Graduate Research Assistant, Department of Civil Engineering, North Carolina State University, Raleigh, NC 27695-7908

[3]President, The Collin Group, 11 Plantation Court, N. Bethesda, MD 20852

Introduction

Shallow foundations are commonly used to support commercial and domestic buildings, storage tanks, bridge abutments, and military installations. Construction and performance of shallow foundations in cases where in situ soil conditions encompass loose or soft soils represent a continuous challenge to the profession.

Construction in loose and soft subsurface soil conditions may lead to unacceptable levels of deformation and excessive stresses on the superstructure. Consequently, a means of soil improvement may be needed. An emerging method of soil improvement for shallow foundation construction is to excavate a certain depth of the loose/soft soil and use mechanically stabilized backfill in which polymeric materials, including geogrids and geotextiles, are used for reinforcement. Using proper construction and installation, the merit of using polymeric reinforcement is to reduce the amount of deformation under the applied load by dissipating the stress to a level that can be sustained by the underlying soft soil. Therefore, improved load-deformation characteristics can be realized in terms of reduced total and differential settlement.

The evaluation of the stress distribution with depth due to surface loads for shallow foundation is a major requirement for the computation of the foundation' settlement. Presently, methods based on elastic solutions are used in practice. These methods which provide the distribution of the stresses in the vertical and horizontal directions, are relatively easy to use, and with some manipulation, they can be applied to complex foundation layout. However, there is a dearth of information regarding the adequacy of the elastic methods for predicting the stress distribution in a reinforced soil mass.

Work conducted in this research aims at measuring the stress distribution with depth during a plate load testing program on geogrid-reinforced sand. A total of five load tests are performed; four on geogrid-reinforced sand and one on unreinforced sand. The tests are performed in a 1.52m x 1.52m x 1.37m(length x width x depth) steel box with plate dimensions of 0.33 m x 0.33 m. Three layers of geogrid reinforcement are consistently used per reinforced sample with various spacing between the top layer and the test plate. The pressure distribution with depth is measured using six pressure cells; three are located at 0.33 m and three are located and 0.67 m below the test plate. The three pressure cells per layer consist of two 229 mm diameter cells placed on the outside of the plate boundaries and one 50 mm diameter cell that is placed directly under the center of the plate. The testing procedure regarding load increments and duration of loading is in conformance with ASTM 1194-95. A comparison between the stress distribution in the reinforced and unreinforced sand is presented along with a demonstration of the applicability of the elastic methods to the measured data.

Background

Evaluation of the stress distribution with depth due to surface loads is a major requirement for the design of shallow foundation and for the computation of the foundation' settlement. Traditionally, and as presented in many soil mechanics text books, for example Holtz and Kovacs (1981) and Das (1993), the stress distribution in unreinforced soils is evaluated using one of the following methods:

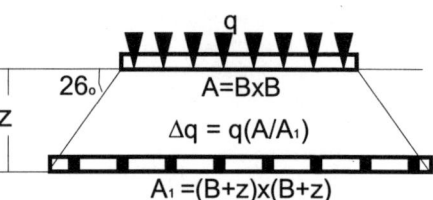

Figure 1. Simplified Load Distribution Scheme

i. Approximate Methods (simplified load spreading, Figure 1): in this case, the angle of propagation of the stresses from the foundation bearing level through the subsurface is assumed based on the soil type. A value commonly used in practice is 26° (which corresponds to 2:1 distribution.)

ii. Elastic Methods: a collection of these methods is presented in the text book by Poulos and Davis (1974). These methods are based on assuming that the soil behaves as a linear elastic medium. Original developments were carried out by Boussinesq and Westergaard with the latter presenting solutions for elastic soil embedded with alternating stiff layers. Elastic solutions for layered soil profiles and for soil profiles with rigid boundaries were later developed by Burmister (1956, 1962) and others.

iii. Numerical Methods: these methods are reserved for complicated foundation configurations or when it becomes necessary to incorporate soil non-linearity in the solution. In this case, the stress distribution and the settlement can be evaluated using the finite element method, the boundary element method, and/or the finite difference method.

There are advantages and disadvantages for each of the above categories of stress-evaluation methods. A main advantage of the approximate method is its simplicity and ease of use. An obvious disadvantage of this method is the lack of accuracy and the nature of the approximate results. Only average stresses can be evaluated at a given depth with no distinction given to the location at which these stresses are being evaluated under the loaded foundation area. In addition, the stress magnitude outside the zone of influence, defined by the distribution angle from the bearing level, is assumed to be zero.

The elastic methods provide the distribution of the stresses in the vertical and horizontal directions and are relatively easy to use. With some manipulation they can be used for complex foundation layout. A main disadvantage of these methods is the inability to account for the soil non-linearity and model soil irregularities. However, this category of methods is the one mostly used in today's state of practice.

The numerical methods are the most versatile and address the shortcomings of the other methods. With the advent of the computer technology, the use of the numerical methods has become more accessible to the engineering community. However, in addition to being relatively expensive and time consuming to apply, these methods are prohibitive in cases where foundation configurations necessitate 3-D modeling. In this case, few programs in the geotechnical arena are deemed suitable and even then the inability to estimate the proper input soil data renders the use of these programs in the state-of-practice questionable.

Past studies reported in literature, as presented for example by Guido et al (1985,1987), Das (1994a,b), Yetimoglu et al (1994), Ismail and Raymond (1995 a,b) and Adam and Collin (1996), reported an increase in the bearing capacity of the foundation when geogrid-reinforcement was included. In several instances, results from past studies indicated the presence of a critical u/B ratio (u=spacing between foundation and first layer of reinforcement and B=foundation width) for which the maximum increase in the BCR (defined as bearing capacity of reinforced foundation/bearing capacity of unreinforced foundation) was obtained. This u/B ratio was estimated to be between 0.25 and 0.75 depending on the number of reinforcement layers, spacing, and reinforcement stiffness.

Experimental Program

A total of five plate load tests were performed on sand. A steel test box was constructed for the performance of the plate load tests under controlled conditions. The dimensions of the box were selected to be 1.52m x 1.52m x 1.37m based on preliminary stress distribution analysis to ensure a minimal interference from the box boundaries on the test results. Three wire strain gages were mounted along the side of the box to monitor the occurrence of any induced strain during the application of the load increments. These gages did not register any increasing strain magnitude during loading. The test plate was 0.33m x 0.33m square plate and the size of the soil samples was 1.52m x 1.52m x 1.22m. The box setup is presented in Figure 2.

The load increments were applied to the test plate using a hydraulic jack with 71 kN capacity. The load was applied in increments with each increment maintained until settlement rate was less than 0.05 mm per hour. The magnitude of a given pressure increment at the surface was monitored using an electronic load cell. The testing procedure regarding load increments and duration of loading was in conformance with ASTM 1194 (1995) " Bearing Capacity of Soil for Static Load and Spread Footings".

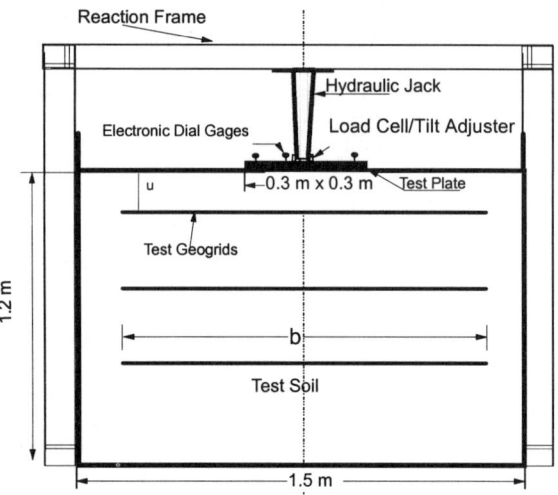

Figure 2. Test Box and Configuration

The pressure distribution with depth was measured using six Model 4800 Geokon earth pressure cells; three were located at 0.3 m and three were located and 0.6 m below the test plate. The three pressure cells per layer consisted of two 228 mm diameter cells placed on the outside of the plate boundaries and one 50 mm diameter cell that was placed directly under the center of the plate. The pressure cells consisted of circular stainless steel plates that were welded together around their periphery and spaced apart by a narrow cavity filled with antifreeze solution. High-pressure stainless steel tubing connects the cavity to a pressure transducer. External pressure acting on the cells is balanced by an equal pressure induced in the internal fluid. This pressure is converted by the pressure transducer into an electronic signal that is transmitted by a four-conductor shielded cable (direct burial type) to the Vibrating-Wire readout box (Geokon Incorporated, 1996).

Test Soil

The test sand known as "Ohio River sand" is typically used in mortar and concrete mixes. Grain size analysis (ASTM D 422-95), specific gravity tests (ASTM D 854-95), and relative density tests (ASTM 4253 and 4254) were performed on sand specimens in accordance with ASTM Standards (1995). Results from these tests indicated that the sand was uniformly graded with an average specific gravity of 2.7, a maximum dry unit weight of 18.8 kN/m^3 and a minimum dry unit weight of 15.7kN/m^3. The grain size distribution of the test sand is shown in Figure 3. The sand contained less than 3% finer than #200 sieve and has a coefficient of uniformity (c_u) equal to 8 and a coefficient of curvature (c_c) equal to 1.0, indicating a well graded

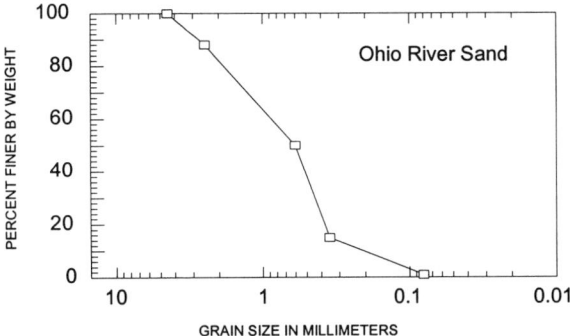
Figure 3. Grain Size Distribution of Test Soil

sand. This sand is classified as SW according to the Unified soil Classification system (USCS).

Test Geogrid

Two grades of biaxial geogrids were utilized in the testing program. The biaxial geogrids were made from polypropylene with apertures of 25 mm and 33 mm in the machine and cross machine directions, respectively, and have a minimum resin content of 99%. The carbon black content was 1%. The geogrids used in testing were trimmed to the size of 1.49m x 1.49m. The tensile strength in the machine and cross machine directions, respectively, was 204 and 292 kN/m for GR1 and 270 and 438 kN/m for GR2.

Sample Preparation Technique

Figure 4 (a,b) shows the lift thickness configuration and the locations of the pressure cells for a top geogrid spacing of 150 mm and 300 mm, respectively, with a uniform spacing of 300 mm. The locations of the pressure cells are also marked on the figure. The number of geogrids for all of the reinforced tests was 3.

The soil was compacted in the test box in lifts using a jackhammer fitted with a 203 mm x 203 mm tamping plate. The jackhammer delivered approximately 40 mm-kN blows at the rate of 1100 blow/minute. The compaction commenced in one corner and proceeded to the other corner while staying on each plate footprint for

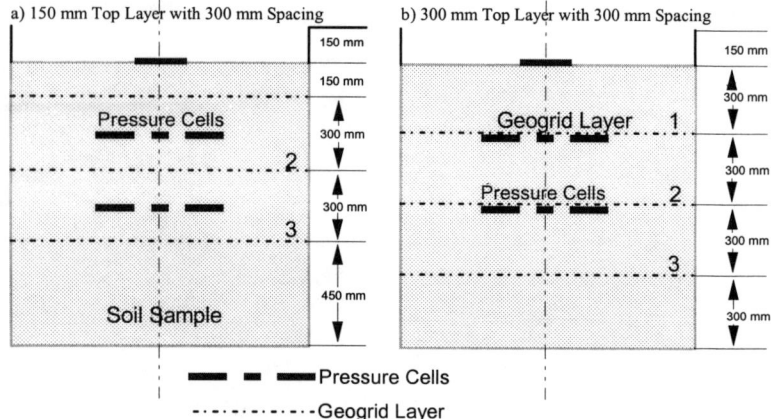

Figure 4. Configuration and Test Spacing

approximately five seconds. This process was repeated until the entire lift surface was uniformly compacted. After the completion of compaction, a layer of geogrid was installed and the next lift was consequently prepared.

The nuclear density/moisture gage was used to measure the moisture content and density distribution per lift according to ASTM D2922-95 for density and D3017-95 for moisture content. The nuclear gage was oriented in the long direction with its sides parallel to the box's sides and the nuclear moisture/density tests were performed for a duration of one minute in the direct transmission mode. Five tests per lift were performed; at the four corners and at the center of the soil sample. The unit weight achieved for the tests ranged from 17.3 kN/m^3 to 17.9 kN/m^3 (relative density on the order of 60%) with water content that ranged from 1.00% to 2.50%. Results of triaxial tests on the sand yielded an effective friction angle equal to 38.6°.

Testing Program

Table 1 shows the number of tests and test configuration used in the experimental program. A total of five tests were conducted on the sand samples. Two tests with the GR1 grids were performed to estimate the effect of the spacing of the first layer (152 mm and 305 mm were used in this study) on attenuating the stresses. One test was performed using GR2 geogrids with u=229 mm to discern the effect of the geogrid stiffness.

Table 1. Testing Program: Number of Tests and Geogrids Spacing

	Reinforcement Spacing				No Reinforcement
	GR1			GR2	
Top Layer Spacing	152 mm[1]	305 mm[1]	229 mm[2]	229 mm[2]	
u/B ratio	0.5	1	0.75	0.75	
Sand	1	1	1	1	1

[1] Bottom layers uniform spacing =305 mm
[2] Bottom layers uniform spacing =229 mm

Pressure Distribution

The measured pressure distributions for the five tests are shown in Figure5 (a,b,c). The results are presented in terms of stress influence factor (I) which is defined as the measured stress at a given depth divided by the surface pressure. The stresses measured by the pressure cells are the total stresses induced under the center of the test plate.

As apparent in Figure 5 (a,b,c), the stress influence factor (I) is a function of the load level and increases as the loading level is increased. This behavior may be expected as the stress-strain characteristics of the test soil are non-linear and can be attributed to the variation of the soil modulus with the increased loading level. Under the applied surface stress of 229.6 kPa, an I value of 0.55 was obtained at 0.3 m depth for the unreinforced case. In comparison, an I value of 0.25 was obtained for the case of GR2 with 228.6 mm uniform spacing versus 0.35 for the GR1 geogrids with the same spacing. At the same surface stress level and for GR1, reducing the top spacing from 305 mm to 152 mm resulted in reducing the I value from 0.3 to 0.25.

Boussinesq and Westergaard Stress Distributions

The stress distribution under the corner of a rectangular loaded area over semi infinite isotropic elastic mass was given by Newmark (1935) and was obtained by integrating Boussinesq's equation for a point load to obtain the following:

$$\Delta q = \frac{q}{4\pi}\left[\frac{2mn(m^2+n^2+1)^{1/2}}{m^2+n^2+m^2n^2+1}\cdot\frac{m^2+n^2+2}{(m^2+n^2+1)} + \tan^{-1}\frac{2mn(m^2+n^2+1)^{1/2}}{m^2+n^2-m^2n^2+1}\right] \quad (1)$$

where q=applied surface stress, m =B/z and n=L/z, BxL= area dimensions, and z=depth.

In comparison, the stress distribution beneath a rectangular area over a semi-infinite mass reinforced by horizontal strata that prevent the deformation in the horizontal

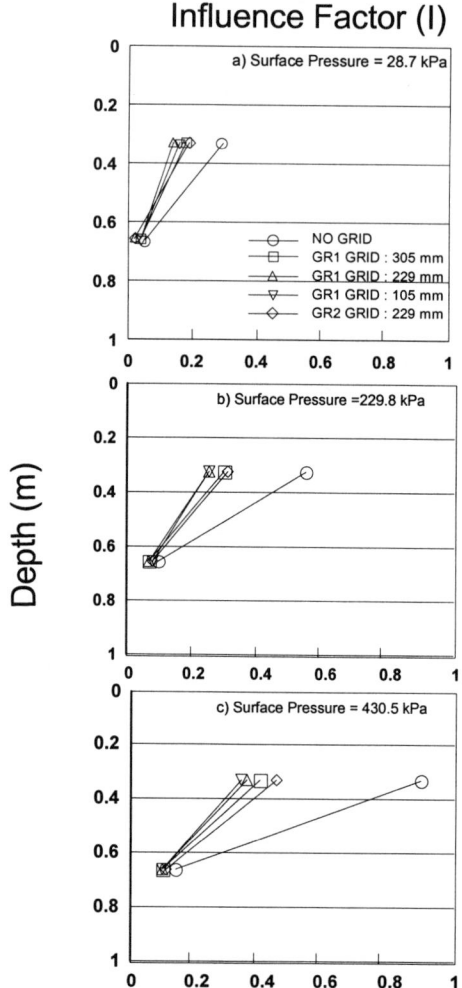

Figure 5: Stress Influence Factors from Measurements as a Function of Applied Stress

direction was given by Fadum (1948) in the form of influence factors ($\Delta q = q.I$). The solution was developed by integrating the following equation by Westergaard (1938) over a rectangular area:

$$\Delta q = \frac{Q}{z^2}\frac{C}{2\pi}\left[\frac{1}{C^2+(r/z)^2}\right]^{3/2} \text{ and}$$

$$C = \sqrt{\frac{1-2\mu}{2(1-\mu)}}$$

(2)

where Q=point load, z=depth of the point of interest at which the stress is evaluated, and r=offset distance of the point of interest from the location of load application. Harrison, and Gerrard (1972) indicated that the Westergaard solution corresponds to the limiting case of horizontal soil modulus=∞.

For the sake of clarity, the distributions of the stress with depth at a relatively low surface pressure (28.7 kPa), medium (229.8 kPa), and relatively high pressure (430.5 kPa) are presented in Figure 6(a,b,c) along with stress estimation from methods based on Boussinesq and Westergaard stress distributions. These stress data indicated a zone of influence that was approximately 1 m (3B).

In comparing measured and computed stresses, the Westergaard stress distribution matched the reinforced data while the Boussinesq distribution was rather conservative under surface stress of 28.7 kPa. On the other hand, and at the relatively high surface stress level of 430.5 kPa, both methods overpredicted the measured data with a closer match with the reinforced data obtained using the Boussinesq distribution. In this case, the measured data were underpredicted by approximately 50% when the Westergaard stress distribution was used. Both methods yielded rather unconservative predictions for the unreinforced case under surface stresses of 430.5 kPa. This may be due to the fact the both of these methods were developed with the assumption of a material having linear stress-strain characteristics which is not the case under relatively high stresses.

Approximate Method

Applying the concept of the approximate stress distribution to these data, the angle of stress propagation (α) can be estimated for the tests on the reinforced sand. Assuming a "2:1" distribution for the unreinforced sand, the angle of the stress distribution for the reinforced sand (α) was estimated for the different tests and presented in Figure 7. The following assumptions were advanced for computing α using the approximate method:

1) The angle α is estimated using the stresses measured under the center of the plate

GEOSYNTHETICS IN FOUNDATION REINFORCEMENT

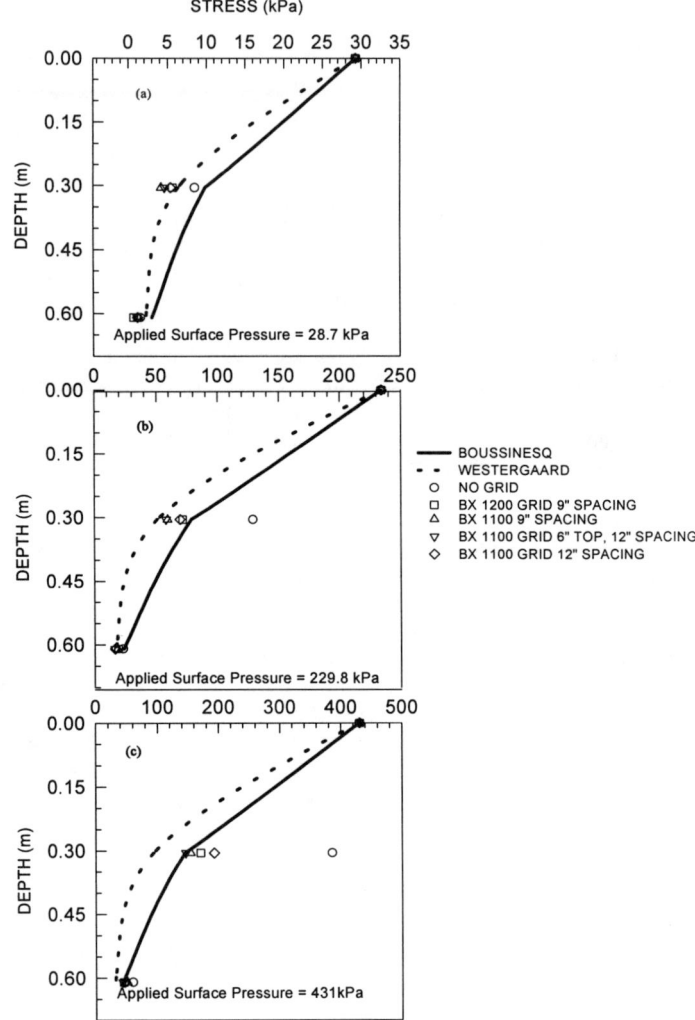

Figure 6: Measured and Predicted Stress Distribution with Depth for Three Surface Stresses

2) The stress magnitude outside the zone of influence delineated by the α values is zero.

As shown in Figure 7, the angle of the stress distribution (α) for the unreinforced case ranged from 23° at surface pressure of 28.7 kPa to 1.6° at surface pressure of 430.5 kPa. In comparison, the angle of the stress distribution (α) increased as the geogrid reinforcement was introduced. Compared to the case of no reinforcement, the largest increase in α was observed in the tests where the GR1 geogrid was used. At the relatively low surface pressure of 28.7 kPa, the α estimated from the GR1 tests was increased from 34° for the 305 mm spacing to an average of 40° when 152 mm top spacing was used. A significant dependancy on the stress level was observed by which the α values decreased with increasing load level.

A smaller α was estimated for the GR1 test with 229 mm uniform spacing. In this case, α decreased from approximately 39.5° for surface stress of 28.7 kPa to approximately 18.5° for surface stress of 430.5 kPa. The test utilizing the GR2 and 229 mm uniform spacing indicated an α value of 32.5° for surface stress of 28.7 kPa and 16.5° for surface stress of 430.5 kPa, respectively.

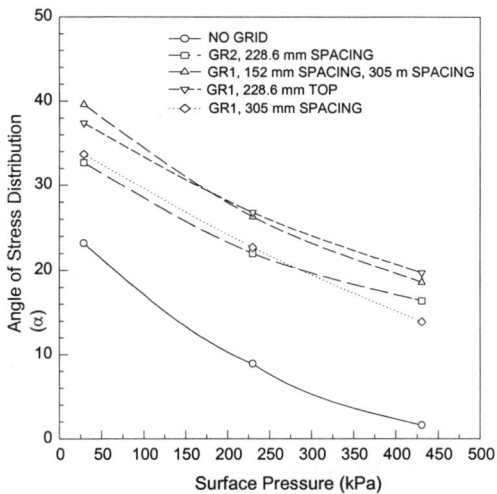

Figure 7: Variation of Angle of Stress Distribution α as a Function of Applied Surface Pressure

Summary and Conclusions

An experimental study was conducted to evaluate the stress distribution in sand reinforced with stiff polymeric geogrids. Two types of geogrids were used; GR1 and GR2 and a total of five plate load tests were performed. Six pressure cells were placed throughout the testing box in two layers to measure the stress distribution with depth. A non-linear stress increase was measured as the applied surface pressure was increased with the consistent trend of lower measured stresses with the inclusion of the geogrids. Based on the results obtained from this study, the following conclusions can be advanced:

1. The magnitude of measured stresses for the reinforced sand was reasonably predicted using the Westergaard method for applied surface pressure of 28.7 kPa and 229.8 kPa. At the high stress of 430.5 kPa, computed data from both Westergaard and Boussinesq distributions overpredicted the measured stresses.

2. The measured data for the unreinforced test were consistently underpredicted using both methods (Boussinesq and Westergaard) when the applied surface stresses exceeded 28.7 kPa.

3. Reducing the data in accordance with the approximate method, higher values of the angle of the stress distribution (α) were estimated for the reinforced sand as compared to the unreinforced samples.

4. In the case of the unreinforced sand, the angle of the stress distribution (α) ranged from approximately 23^0 at surface pressure of 28.7 kPa to 1.5^0 at surface pressure of 430.5 kPa. In comparison, the (α) for the GR1-152mm top spacing case ranged from 40^0 at surface pressure of 28.7 kPa to 19^0 at surface pressure of 430.5 kPa.

5. Lower (α) values were observed as the surface pressure was increased for both the reinforced and the unreinforced tests. However, the rate of α-reduction was larger for the case of the unreinforced sample.

6. The applicability of the results, and conclusions, presented in this study is limited to conditions similar to those used during the testing program.

The results presented in this study indicated an improvement in the plate load carrying capacity and reduction in the stresses with depth as the geogrids were included. This may be explained by the confining and interface friction effects of the test geogrids. In effect, the inclusion of the geogrid layers transform the compacted soil to a highly anisotropic material with relatively larger modulus, as compared to the unreinforced soil, due to the soil-geogrid interface shear resistance as well as the influence of confinement on the lateral deformation and stress propagation. However, in order to fully understand the load-deformation behavior of reinforced soils, instrumentation of the geogrids should be performed to estimate the stress taken by the geogrids and correlate it to the phenomenon of stress reduction.

Acknowledgment

This study was sponsored by Tensar Earth Technology, Inc. Undergraduate students Robert Dodson, Keith Aragona, and Brent Trimble assisted in preparing the test samples and their help is greatly appreciated. Mr. David Turner provided assistance with the test set up and sample preparation techniques.

References

Adams, Michael T. and James G. Collin (1997) "Large Model Spread Footing Load Tests on Geosynthetic Reinforced Soil Foundations," Journal of Geotechnical and Geoenvironmental Engineering, ASCE, Vol. 123 No. 1 pp 66-72 .

American Society for Testing and Materials (1995) Annual Book of ASTM Standards, Section 4, Volume 4.08, Soils and Rock, Philadelphia, PA.

ASTM (1995) Annual Book of Standards, D2922, Standard Test Method for Density of Soil and Soil-Aggregate in Place by Nuclear Method (Shallow Depth) Volume 4.08, Soils and Rock (I), Philadelphia, PA, pp. 257-261.

ASTM (1995) Annual Book of Standards, D3017, Standard Test Method for Water Content of Soil and Rock in Place by Nuclear Methods (Shallow Depth) Volume 4.08, Soils and Rock (I), Philadelphia, PA, pp. 285-289.

Burmister, D.M. (1956) "Stress and Displacement Characteristics of a Two-Layer Rigid Base Soil System: Influence Diagrams and Practical Applications," Proceedings, Highway Research Board, Vol 35, pp. 773-814

Burmister, D.M. (1962) "Application of Layered System Concepts and Principles to Interpretations and Evaluations of Asphalt Pavement Performance and to Design and Construction," Proceedings, International Conference on Structural Design of Pavements, University of Michigan, Ann Arbor, Michigan, pp. 441- 453

Das. B.M. !993) Principles of Heotechnical Engineering. PWS Publishing Company, Boston, MA.

Fadum, R.E. (1948) "Influence Values for Estimating Stresses in Elastic Foundations,"Proceedings, 2nd International Conference on Soil Mechanics and Foundation Engineering, Vol. 3, pp 77-84.

Geokon Incorporated (1996) "Instructional Manual Model GK-403, VW Earth Pressure Cells," Geokon Geotechnical Instrumentation, Lebanon, NH.

Guido V. A., Biesaidechi G. L., and Sullivan M. J. (1985) "Bearing Capacity of a Geotextile Reinforced Foundation," Proceedings, 11th International Conference on Soil Mechanics and Foundation Engineering, San Francisco, Vol. 3, pp 1777-1780.

Guido V. A. , Knueppel J. D. and Sweeny M. A. (1987) " Plate Loading Tests on Geogrid Reinforced Earth Slabs," Proceedings, Geosynthetics '87, New Orleans, pp. 216-225.

Harrison, W. J., and Gerrard, C. M. (1972) "Elastic Theory Applied to Reinforced Earth," Journal of Soil Mechanics and Foundation Division, ASCE, Vol 98, No. Sm12, pp. 1325-1345.

Holtz, R.D., and Kovacs, W.D. (1981) An Introduction to Geotechnical Engineering. Prentice Hall Civil Engineering and Engineering Mechanics Series, Englewood Cliffs, N.J.

Ismail I., and Raymond G. P. (1995a) " Investigation Reveals the Interface is not the Best Place for Geosynthetic Reinforcement," Part I, Geosynthetic Fabrics Report, May 95, pp. 15-19.

Ismail I., and Raymond G. P. (1995b) " Investigation Reveals the Interface is not the Best Place for Geosynthetic Reinforcement," Part I, Geosynthetic Fabrics Report, May 95, pp. 9-14.

Newmark, N. M. (1935) " Simplified Computation of Vertical Pressures in Elastic Foundations," University of Illinois Engineering Experiment Station, Circular No. 24.

Poulos, H. G., and Davis E. H. (1974) "Elastic Solutions For Soil and Rock Mechanics," John Wiley and Sons, New York, pp 411.

Yetimoglu, T. , Wu, J. T. H., and Saglamer, A. (1994) " Bearing Capacity of Rectangular Footings on Geogrid-Reinforced Sand," Journal of Geotechnical Engineering, ASCE, Vol. 120(12), pp. 2083-2099.

GEOSYNTHETIC EROSION CONTROL MATERIALS: A LANDFILL COVER FIELD STUDY

George R. Koerner
Geosynthetic Institute, Drexel University, Philadelphia, PA, U. S. A.

David A. Carson
U. S. Environmental Protection Agency, Cincinnati, OH, U. S. A.

ABSTRACT

This paper presents information gathered as part of a study to examine landfill slope stability. Geosynthetic erosion control materials were placed on the surface of thirteen field test plots at two slope angles commonly found in landfill applications: i.e., 2H :1 V (26.6 degrees) and 3H :1 V (18.4 degrees). A control plot was constructed without any erosion control material. This study consists of visual observations vis-a-vis the respective plot . The general effectiveness of the geosynthetic erosion control materials are summarized over four years of service.

BACKGROUND

Excessive sediment in waterways is a national and worldwide problem. The Clean Water Act of 1972 and the Department of Agriculture's Farm Bill mandate against excessive soil erosion. Both natural and man-made erosion control materials have resulted in significant improvements in mitigating soil erosion. The goal of any revegetation or erosion control project should be to stabilize soils and control erosion in an economical manner.

Soil erosion involves the detachment, transportation and deposition of soil particles by water or air. On bare soil, the movement of water and/or air begins the erosion process by detaching soil particles. These particles collide to displace other soil particles thus increasing the erosive force down the slope. The soil/water or soil/air mixture then begins to downcut the slope, forming erosion sheets or rills. These sheets or rills begin to channelize flow forming gullies. The formation of gullies may eventually lead to successive or even catastrophic failure of the slope. Solutions typically involve the use of erosion control materials which keeps the water and wind from eroding the soil particles. Such erosion control materials generally provide soil cover as well as a surface structural substrate which provides some degree of soil retention, and can facilitate the establishment and/or reinforce vegetation.

Final covers of closed landfills present a unique erosion control challenge. As noted by Harris, Rivette and Spradley (1992) soil loss results in greatly increased operational and closure costs for landfill operators. In addition to financial impact, studies have shown that runoff from construction sites can contain organics, nutrients and heavy metals. This has taken on added significance since the United States Environmental Protection Agency (U.S. EPA) developed the Non Point-source Discharge Elimination System (NPDES) Permit Application Regulations for Storm Water Discharges in 1992. The need for proper erosion control from sloped surfaces is obvious.

In late 1994, a research effort was focused on full scale test plots to assess the internal stability of geosynthetic clay liners (GCLs) when placed on steep slopes, Carson et. al. (1998). Since the surface of the test plots had to be protected against erosion, the opportunity presented itself to study erosion control materials. The benefits realized by utilization of erosion control materials on this project have been qualitative rather than quantitative in nature. There is a need to collect field data from actual case histories so that findings can be evaluated. The objective of this effort is to ascertain the performance of a wide variety of erosion control test results from a single field site. In so doing, technical improvements in geosynthetic erosion control materials as well as a greater awareness of erosion control materials may be realized.

SITE LAYOUT

The field test consists of fourteen test plots, as shown in plan view in Figure 1. The study is ongoing, thus this paper represents approximately 4 years of evaluation. Nine of the test plots were constructed on 2:1 slopes (plots F, G, H, I, J, K, L, M, N, and P) and five were constructed on 3:1 slopes (plots A, B, C, D, and E). Plot M was installed on the 2:1 slope to act as the control section for the various erosion control materials placed on all of the other plots. In the case of plot M, no geosynthetic erosion control material was installed. Plot P is positioned where G and H used to reside.

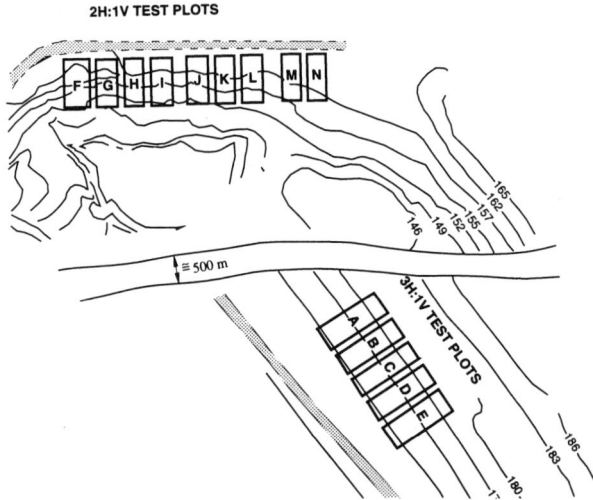

Figure 1, Plan View Layout of Test Plots (contours are in units of meters)

The length of the plots range from 20 to 29 m. The width of the plots varied between 6 and 9 m. The targeted slope angle for the 2:1 slopes was 26.6 degrees and for the 3:1 slopes was 18.4 degrees. In actuality the slope angles for the 2:1 slopes ranged from 22.9 to 25.5 degrees. The slope angles for the 3:1 slopes ranged from 16.9 and 17.8 degrees. Figure 2, shows a typical cross sections for the test plots containing GCLs. As mentioned , test plot M was the erosion control section and contained no GCL (nor other geosynthetics).

GEOSYNTHETIC EROSION CONTROL MATERIALS (GECM)

The geosynthetic erosion control materials used for this project were generously donated by three companies; AKZO Nobel Geosynthetics, Synthetic Industries, Inc. and The Tenser Corporation. Their contribution of this material was greatly appreciated. Table 1, modified from Daniel and Scranton (1996), presents a schedule of the geosynthetic erosion control materials used on this project. In addition, Table 2 presents minimum average role value (MARV) product data which characterizes the geosynthetic erosion control materials physically and mechanically. Note that data presented in Table 2 is from the manufacturers literature and in minimum average roll values of the material.

Table 2 also contains typical unit costs for the various erosion control materials. These unit costs are installed costs in the units of U.S. dollars per square meter of material. They are estimates only and were obtained from conversations with the manufacturers. They represent costs at the time of installation. These cost are both project specific and location specific and change over time. Hence they should only be used on a comparative basis for this study.

EROSION CONTROL CONSTRUCTION SEQUENCE

The installation and construction of the geosynthetic erosion control materials occurred over a two week period at the end of November, 1994. Initially, cover soil was placed over the test plots with a low contact pressure dozer pushing soil up the slope and a long arm hydraulic hoe from the top of the slope. Upon making final grade, the toe of each plot was removed with a track mounted back hoe. Soil densities were measured after placement.

The extent of soil preparation was "tracking" the final cover soil lift with the dozer. This resulted in a washboard type surface. The estimated in situ conditions of the upper lift of soil are as follows;

Soil classification = SM
agronomy classification = sandy clay loam
average approximate density = 1762 kg/m^3
average approximate moisture content = 30%

The placement of the erosion control materials began at the top of the slope. The erosion control blanket was deployed from the top of the slope

Table 1 - Erosion Control Material Schedule for the EPA Landfill Cover Test Plots in Cincinnati, Ohio

Plot	Slope	Manufacturer and Product	Type	Roll Width (m)	Roll Length (m)	Composition	Color	Manufacturer's Suggested Application
A	3H:1V	Tensar TB1000	ECB	2.3	23	polyolefin	green	steep slopes
B	3H:1V	Syn. Ind. Landlok® 407 GT	ECB	3.8	132	degradable polypropylene	beige	slopes, low flow ditches
C	3H:1V	Syn. Inc. Landlok® 407 GT & Excelsior	ECB	3.8	132	degradable polypropylene	beige	slopes, low flow ditches
D	3H:1V	Akzo Enkamat 7010	TRM	1	152	nylon	black	low velocity slopes
E	3H:1V	Akzo Enkamat 7010 & Excelsior	TRM	1	152	nylon	black	low velocity slopes
F	2H:1V	Tensar TM3000	TRM	1.5	31	polyethylene	black	high flow channels
G	2H:1V	Tensar TM3000	TRM	1.5	31	polyethylene	black	high flow channels
H	2H:1V	Tensar TM3000 & Excelsior	TRM	1.5	31	polyethylene	black	high flow channels
I	2H:1V	Syn. Ind. Landlok 450 & Excelsior	TRM	2	42	polyolefin	green	ditches steep slopes
J	2H:1V	Syn. Ind. Landlok 450 & Excelsior	TRM	2	42	polyolefin	green	ditches steep slopes
K	2H:1V	Akzo Enkamat 7010 & Excelsior	TRM	1	152	nylon	black	low velocity slopes
L	2H:1V	Akzo Enkamat 7010	TRM	1	152	nylon	black	low velocity slopes
M	2H:1V	none	na	na	na	control	na	na
N	2H:1V	Akzo Enkamat 7010 & Excelsior	TRM	1	152	nylon	black	low velocity slopes
P	2H:1V	Akzo Enkamat 7020	TRM	1	85	nylon	black	high velocity slopes

ECB = erosion control blanket, a temporary degradable rolled erosion control product composed of processed natural or polymer fibers mechanically structurally or chemically bound together to form a continuous matrix.

TRM = turf reinforcement mat, a long term non-degradable rolled erosion control product composed of UV stabilized, non-degradable, synthetic fibers, nettings and/or filaments processed into three dimensional reinforcement matrices designed for permanent and critical hydraulic applications where design discharges exert velocities and shear stresses that exceed the limits of mature, natural vegetation. TRMs provide sufficient thickness, strength and void space to permit soil filling and/or retention and the development of vegetation within the matrix.

Table 2 – Erosion Control Material Properties for the Cincinnati, Ohio Landfill Cover Test Plots

Plot	Manufacturer and Product	Thickness ASTM D5199 (mm)	Mass/Unit Area ASTM D5261 (g/m²)	MD Tensile Strength ASTM D5035 (kN/m)	MD Tensile Strain ASTM D5035 (%)	UV Resistance ASTM D4355 % Strength Retained @ 500 hrs.	Cost $/sq. m
A	Tensar TB1000	10.2	339	6.4	35	80	3
B	Syn. Ind. Landlok 407 GT	1	71	9.6	25	5	1.5
C	Syn. Inc. Landlok 407 GT & Excelsior	1	71	9.6	25	5	2.5
D	Akzo Enkamat 7010	8.9	247	2.2	43	84	4
E	Akzo Enkamat 7010 & Excelsior	8.9	247	2.2	43	84	5
F	Tensar TM3000	12.7	406	7.9	40	80	4
G	Tensar TM3000	12.7	406	7.9	30	80	4
H	Tensar TM3000 & Excelsior	12.7	406	7.9	30	80	5
I	Syn. Ind. Landlok 450 & Excelsior	10.2	339	2.4	50	90	8
J	Syn. Ind. Landlok 450 & Excelsior	10.2	339	2.4	50	90	8
K	Akzo Enkamat 7010 & Excelsior	8.9	247	2.2	43	84	7
L	Akzo Enkamat 7010	8.9	247	2.2	43	84	6
M	none	na	na	na	na	na	0
N	Akzo Enkamat 7010 & Excelsior	8.9	247	2.2	43	84	7
P	Akzo Enkamat 7020	17.2	376	3.0	49	86	8

downward. The mats or blankets were rolled downslope in the direction of the maximum gradient. The edges of the parallel blankets were overlapped in accordance with the manufacturer's instructions, typically 100 to 150 mm. When necessary, the blankets were spliced down the slope, the blankets were placed end over end (shingle style) with approximately 200 mm of the up gradient slope blanket covering the leading end of the down gradient blanket. Figure 3 shows a general schematic of how the erosion control blankets were deployed. GECMs were also placed in drainage swales between test plots so as to minimize edge effects from adjacent plots.

The overlaps were stapled together with galvanized steel staples 50 mm wide and 150 mm long. The general staple pattern is also depicted in Figure 3. The staples are placed 2 m apart on the parallel (to the slope) edges of the blankets (running along the slope) and 4 to 5 m apart on the perpendicular (running across the slope) edges of the blankets. The staples were positioned by hand and then stomped into placed by foot. In some cases, the staples were too weak to be inserted into the soil which caused delay.

Upon placing the geosynthetic erosion control materials (GECM), some test plots received a layer of excelsior over top of the (GECM). Excelsior blankets are temporary erosion control materials comprised of wood shavings used to increase water retention . Such a characteristic of this "natural" material is believed to aid in the germination of grass seed. Check the plot schedule of Tables 1 and 2 which identifies the plots where excelsior was used.

After placement of the GECM, hydraulic seeding/mulching began. Selection of the type of seed and mulch was based on the following factors;

- time of year (November is the end of the fall grass growing season)
- soil texture-type (silty sandy loam)
- estimate soil moisture regime for site (site is considered normal rather than wet or dry)

The site is considered to have low to medium vegetation maintenance plan. This means that the vegetation is considered more functional than aesthetic. Such a low to medium rating implies that the area receives little if any supplemental fertilization or irrigation and does not require mowing. The project was located in Cincinnati, OH which is on the border of grass adaptation zones 1 and 2, as described by Theisen and Carroll (1990). The regional climate and biological associations of this area are excellent for grass germination.

It should be noted that there is a clear distinction between an erosion control blanket (ECB) and a turf reinforcement mat (TRM). An ECB is, generally a lofty web of fibers trapped between two nets and bound together in spots thermally or via rows of parallel stitches. The material is relatively inexpensive. It has an installed cost of 1.00 to 3.00 dollars a square meter. It is different from a TRM which is a three dimensional structure which is covered with soil. TRMs have greater physical and mechanical properties than ECBs as a result of greater

GEOSYNTHETICS IN FOUNDATION REINFORCEMENT

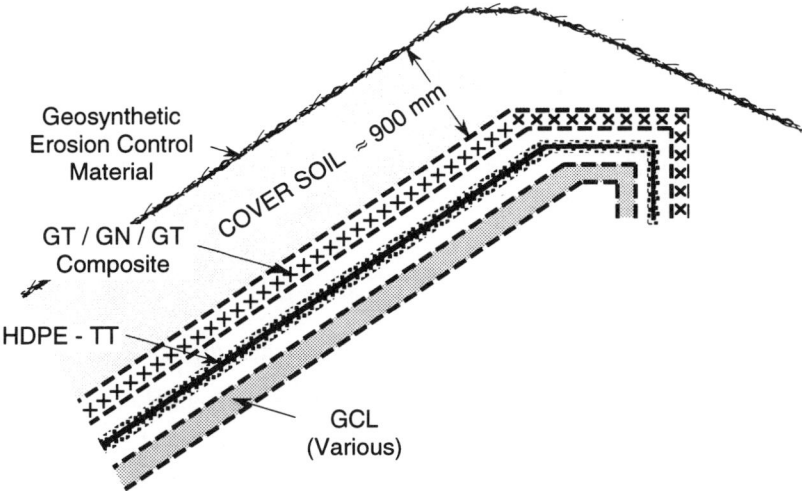

Figure 2, Cross section of typical test plot showing location of geosynthetic erosion control material

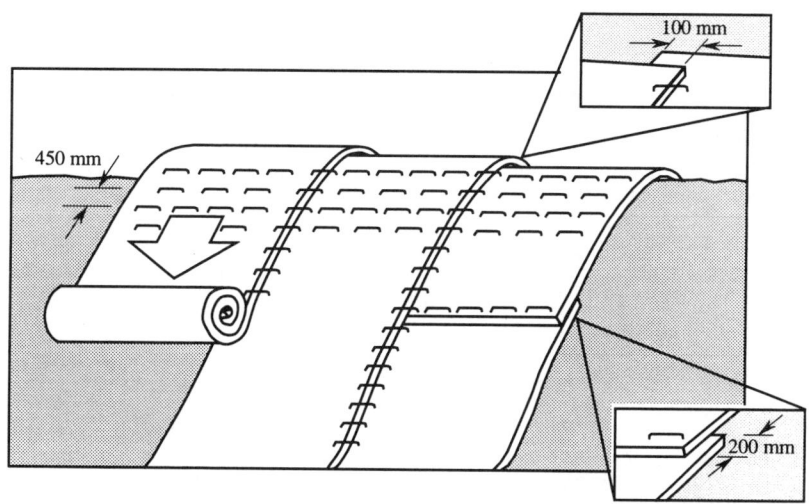

Figure 3, Schematic diagram of installation of geosynthetic erosion control material at the Cincinnati, OH test site

polymer utilization. TRMs are recommended for use in high flow channels or high velocity slopes where long term design velocities are greater than 1.5 m/sec. ECBs were only used on the 3:1 slopes which is the upper bound of their manufacturer's suggested usage.

Hydraulic seeding/mulching was some what complicated. The material applied to the slope consisted of three parts, seed, fertilizer and mulch. The hydraulic seeding/mulching was applied in one of three sequences in accordance with the manufacturer's instructions:

- Seed and fertilizer mix, the Geosynthetic Erosion Control Material (GECM), then hydraulic mulch.

- Hydraulic seed, fertilizer and mulch mix, then GECM.

- GECM, then hydraulic seed, fertilizer and mulch mix.

Germination of the grass seed took 7 to 10 days and fortunately there was three 12 mm rain events within this 10 day time window. In addition, the site experienced a long, mild winter. Therefore severe weather was not realized until late December. Since the erosion control materials were part of an ongoing experiment, maintenance of the vegetation was kept to a minimum. Hand maintenance was occasionally performed, no mowing of the site was allowed.

SITE CLIMATOLOGIC DATA

The climate of the greater Cincinnati area is continental with a rather wide range of temperatures from winter to summer. A precipitation maximum occurs during winter and spring with a late summer and fall minimum. Summers are warm and rather humid with freeze free periods lasting on average 187 days from mid-April to the latter part of October. The heaviest precipitation, as well as the precipitation with the longest duration, is normally associated with low pressure disturbances moving southwest to northeast through the Ohio River valley.

Figures 4 (a) and (b) show the precipitation and average temperature at the project site respectively for the last two years. As can be seen from the two graphs, 1995 and 1996 were rather typical climatic years for the greater Cincinnati area.

PERFORMANCE OBSERVATIONS

The function of geosynthetic erosion control materials (GECM) is to enhance the erosion resistance of vegetated grass slopes. This enhancement is afforded to the slope by the GECM by surface protection and shielding of underlying soil and reinforcement provided to the root structure of the vegetation. This results in dissipation of raindrop energy and minimization of lateral erosion flow.

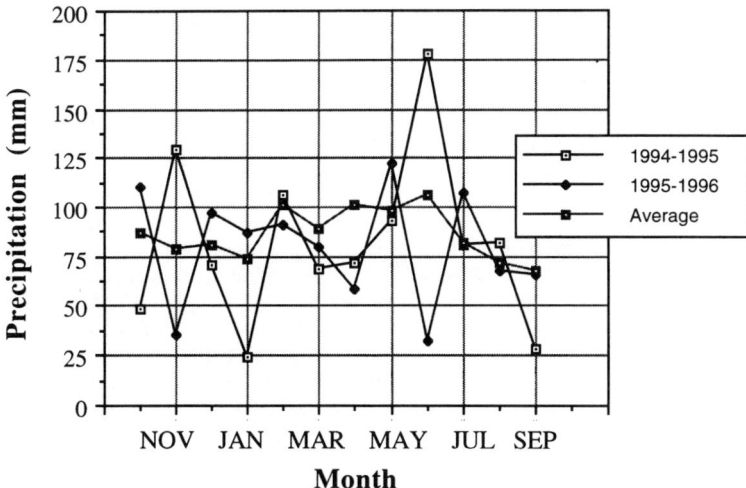

Figure 4 a, First two years of precipitation at the field site

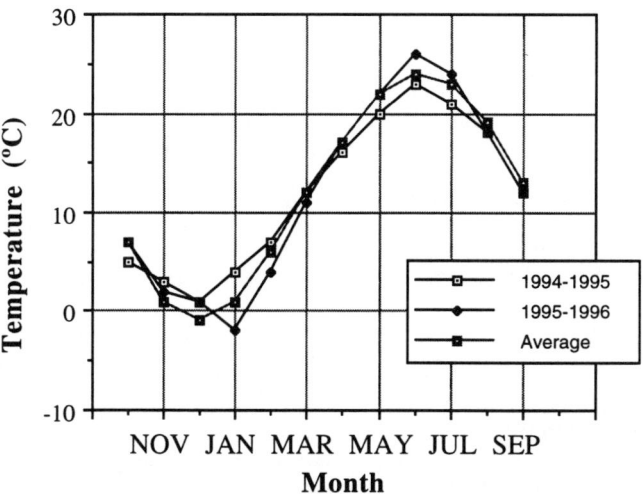

Figure 4 b, Average monthly temperature at the field site over the first two years of service

Since no monitoring devices where used to quantify the performance of the test plots, observations were used to qualitatively assess the various plot performance. Observations were collected over the service time period of four years. These observations were based on ; vegetation established, prevention of surface erosion features (rills) and prevention of toe sloughing. Observations were compared to each other as well as the control plot in which no GECM was used.

Each group of test plots had a control section associated with it. The performance of the control section on the 3:1 slope showed modest erosion features such as rills and slight soil loss. This control section was constructed by tracking a dozer up and down the slope next to plot "A". In hind sight, it would have been better to build a control section from scratch with fill as was done on the control section of 2:1 slope. By only tracking an established grass slope, it is felt that the root system was only disturbed and not total destroyed. Hence the grass of the 3:1 control section grew in fast and thick since the subsoil was not significantly disturbed.

In the cases where an ECB was used alone on the 3:1 slope, (i.e., plots A and C) they worked well and resulted in no observable rill formation. When excelsior was used in conjunction with the ECB the plot performance did not improve. This can be seen by the presence of less vegetation in plot "E" than in plot "D" as can be seen in Figure 5. As noted earlier, construction commenced in the last week of November. Due to this time of year being the end of the growing season, it appears than the grass needed heat rather moisture retention to expedite germination. The excelsior blanket even may have covered too much soil which inhibited germination. On a more positive note, neither the ECB or the ECB and excelsior appear to relinquish much soil. However, the added excelsior showed little benefit. This probably would not have been the case if the plots were built at a different time of year (i.e. summer) or in a different climate.

In the cases where the TRM were utilized on plots "D" and "E" on the 3:1 slope, it appeared that this level of protection was not necessary for this relatively short 30 m slope. Plot "D" and plots "E" appeared in good shape throughout the study but so did the ECB covered slope at less than half the cost.

The performance of the control section on the 2:1 slope showed dramatic signs of distress after only four months of exposure. As can be seen in Figures 6 and 7 the upper section of the "M" control plot showed signs of sheet runoff concentrating and tending to form rills and small gullies. Such features form channels for concentrated flow. This flow becomes progressively more severe as the erosive force of the flowing water increases with slope length (i.e. the further one progresses down the slope the worse the condition becomes). Figure 8 shows the lower half of the "M" control section. Clearly two shallow circular scarps are visible about two thirds of the way down the slope. These scarps are indicative of slope stability failures. This is an example of an extreme case of erosion which is not acceptable. The control plot "M" was in very bad condition after only four months of exposure to a relatively average Cincinnati, Ohio

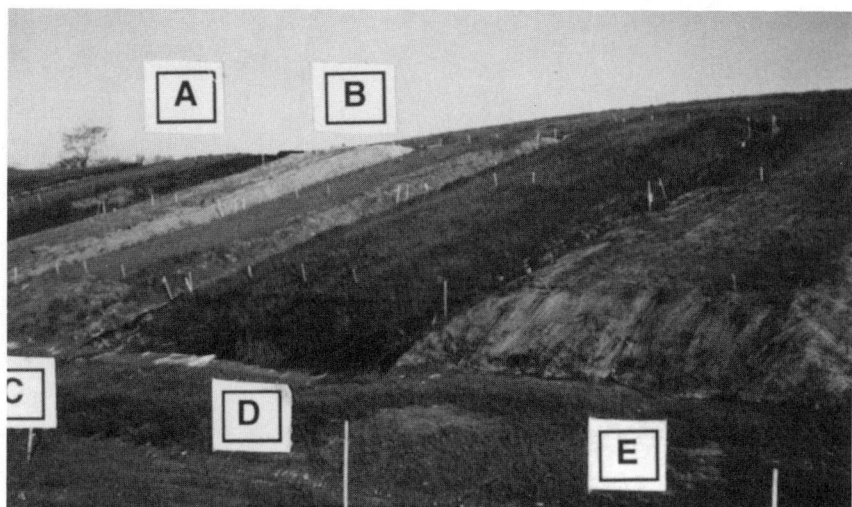

Figure 5, Overview Photograph of 3H:1V test plots A, B, C, D, and E after 4 months of service.

Figure 6, Overview Photograph of 2H:1V test plots F, G, H, I, J, K, L, M, and N after 4 months of service. Plots G and H are unusable for erosion control assessment due to cover system slides at these locations.

Figure 7,   Photograph of the upper portion of the control test plot "M" after 4 months of service. Rills and Gully are clearly apparent.

Figure 8,   Photograph of the lower portion of the control test plot "M" after 4 months of service. Two shallow slip circles are apparent and the toe of the slope has washed out resulting in significant soil loss.

weather year. Furthermore, the toe of the plot has washed out, resulting in significant soil lose.

Due to several instabilities of the cover systems at three of the test plots on the 2:1 slope our experimental design was disrupted. Instability was encountered at plots F, G and H, Koerner et. al. (1996). Plots G and H underwent dramatic slides which involved complete translation of the cover. Plot F experienced a prolonged instability which involved the toe of the slope sloughing away over a period of 5 months. This gradual instability was interesting from a GECM perspective in that it showed that GECMs do not have enough tensile strength to resist a slide of this nature. As the slide progressed, the GECM worked itself towards the surface shedding its cover soil. When the GECM was no longer in contact with the ground surface, deterioration of the plot's toe began to accelerate as forecasted by Rustom and Weggel (1990).

It is important to note that 2:1 slopes are significantly different than 3:1 slopes in regards to the type of erosion protection required. After observing these tests plots for four years it became clear why different GECMs are appropriate for different slopes based solely on slope angle. ECBs are appropriate for 3:1 slopes and TRMs are needed for 2:1 slopes.

GECMs performed well on the 2:1 test plots over the long term. However, over the short term, problems with seed germination on the plots that utilized excelsior were encountered. Due to poor grass growth on nearly all of the 2:1 plots it was decided to reapply hydraulic seed and mulch to all of the 2:1 plots in the Spring of 1995.

SUMMARY

The test plots described in this paper were primarily built to assess the internal stability of geosynthetic clay liners. GECMs were needed to cover these full scale test plots to make the cover simulations realistic. GECM performance was not the primary focus of the study. Without quantitative measures, (such as collecting runoff and the accompanying sediment) and due to external factors such as construction sequencing, reapplication of seed and mulch, and limited resources it is difficult for us to differentiated between the performance of the four different TRMs used as GECMs on the 2:1 slopes. Furthermore, it was difficult to differentiate between Akzo's Enkamat 7010 and Enkamat 7020. The only difference between the two products is that one is half the thickness of the other. Perhaps the shear forces never became high enough to challenge even the thicker product.

CONCLUSIONS

The geosynthetic erosion control materials as a group did an excellent job of controlling erosion for 4 years of this on going study. Ten different erosion control materials were tested on two different slope angles and compared to control sections which failed prematurely.

The following conclusions are based on the results and impressions of the field observation over the past four years.

- Landfill slopes like the ones at the Cincinnati, Ohio GCL test plots benefited in the short term by the rapid establishment of vegetative cover, assisted by GECMs.

- GECMs appear to improve the long term stability of steep and long landfill cover slopes.

- It appears that TRMs are more suitable to steeper slopes, (i.e., 2:1) and ECB may be successfully utilized on more shallow slopes, (i.e., 3:1) .

- The use of excelsior blanket did not appear to improve the performance of the GECMs at either the 3:1 or 2:1 slopes. Perhaps this was due to the season in which the test plots were constructed, the climate of Cincinnati , Ohio and/or the type of hydraulic seeding/mulching that was conducted.

In as much as erosion control is a worldwide problem, every attempt at its control and mitigation is necessary. Geosynthetic erosion control materials are available and have been shown at this site to be very effective. Yet much remains to be done in so far as testing and research is concerned. Quantification of the performance of parallel situations such as presented herein would have been helpful in this regard. Unfortunately, resources were not available since the focus was in another direction. Yet, the paper serves to show that qualitative results, from visual observations are informative, and that this type of layout could wells serve as the blueprint for other field studies resulting in quantitative data.

Knowing the current trend of sites to use ever increasing amounts of native low permeability cover soil (i.e. 10^{-4} to 10^{-6} cm/sec)in landfill covers, it is envisioned that GECM shall become more significant in future landfill cover designs. This becomes particularly significant knowing that these native low permeability cover soil are prone to erosion and post closure care periods are being extended past thirty years.

ACKNOWLEDGMENTS

The authors wish to thank the US EPA for funding this project via cooperative agreement CR-821448 . In addition, the US EPA volunteered personnel to help with the installation and monitor the site over time. Without this level of involvement on the part of the agency this effort would not have been possible.

Geosynthetic erosion control materials used for this portion of the project were generously donated by AKZO Nobel Geosynthetics, Synthetic Industries, Inc. and Tenser Corporation. Their contribution of material as well as help with the installation was greatly appreciated.

REFERENCES

Carson, D. A., Bonaparte, R., Daniel, D. E. and Koerner, R. M., (1998) Geosynthetic Clay Liners at Field-Scale: Internal Shear Test Progress," Proceeding Sixth International Conference on Geosynthetics, Atlanta, GA, pp. 427-432.

Daniel, D. E. and Scranton, H. B. (1996), *Report of 1995 Workshop on Geosynthetic Clay Liners,* EPA/600/R-96/149, US EPA Office of Research and Development, National Risk Management Research Laboratory, Cincinnati, Ohio, pp. 96.

Harris, J. M., Rivette, C. A. and Spradley, G. V., (1992), "Case Histories of Landfill Erosion Protection Using Geosynthetics," The Journal of Geotextiles and Geomembranes, Elsevier Applied Science, Vol. 11 Nos. 4-6, pp. 573-587.

Koerner, R. M., Carson D. A., Daniel, D. E. and Bonaparte, R, (1996) "Current Status of the Cincinnati GCL Test Plots" Proceedings of GRI-10, Field Performance of Geosynthetics and Geosynthetic Related Systems, Philadelphia, PA, pp. 423-445.

Rustom, R. and Weggel, J. R., (1990), "A Study of Erosion Control Systems: Experimental Results," Proceedings 21st Conference International Erosion Control Association., Steamboat Springs, CO, pp. 239-251.

Theisen, M. S., and Carroll, R. G. Jr., (1990),"Turf Reinforcement - The soft Armor Alternative," Proceedings 21st Conference International Erosion Control Association., Steamboat Springs, CO, pp. 255-270.

PARTNERING GEOSYNTHETICS & VEGETATION FOR EROSION CONTROL

by

Robbin B. Sotir[1], John T. Difini[2] and Andrew F. McKown[3]

ABSTRACT

Soil bioengineering for a steep biotechnically stabilized earthen buttress slope construction is initiated with sound engineering practices, which are then integrated with ecological principles. Soil bioengineering solutions which are designed to protect and restore functionality to systems. This was employed on the Massachusetts Turnpike at Mile 79.3 E.B. A polymeric geogrid was coupled with backfill and vegetation in a method known as vegetated geogrid. The use of tensile inclusions made it possible to construct this highly steepened vegetated earthen buttress to reconstruct and stabilize this section of slope along the highway. The fill area above the face was stabilized using a shallow soil bioengineering method known as live fascine, which offered immediate surface erosion control. Environmental and aesthetic goals were paramount in the development of this project.

The soil bioengineering vegetated geogrid and live fascine techniques are useful in constructing steepened slopes to improve mass stability, reduce required fill volumes, provide surface erosion control as well as valuable habitat and aesthetic benefits. The specific vegetated geogrid treatments use live cut branches that are embedded into the slope face on constructed fill terraces. Similar to the polymeric geogrids, the installed branches offer immediate reinforcement and may be considered as supplemental tensile inclusions. The shallow trench live fascine installations reduce a long slope into a series of short slopes controlling surface soil movement. This Turnpike project was constructed in the Winter/Spring of 1995/1996, and serves to illustrate the benefits of using an interdisciplinary approach to solve slope instability revegetation problems.

1. Principal, Robbin B. Sotir & Assoc.Inc., 434 Villa Rica Rd., Marietta, GA 30064
2. Staff Engineer, Haley & Aldrich, Inc., 465 Medford St., 2200 Boston, MA 02129
3. Vice President, Haley & Aldrich, Inc., 465 Medford St., 2200 Boston, MA 02129

INTRODUCTION

Soil Bioengineering Slope Stabilization Techniques

Soil bioengineering uses mechanical, hydrological, biological, and ecological principles to develop living structures for the stabilization and revegetation of cut and fill slopes. The technology is based on sound engineering practices, a fact that is especially evident on this project, with the use of the polymeric geogrids. Forms of these living erosion control techniques have been used in many parts of the world for centuries including Asia, China, Europe, Canada and in the United States beginning in the 1930's by the US Army Corps of Engineers and by the USDA/NRCS. These techniques have since been incorporated into the USDA's National Resources Conservation Service Engineering Field Handbook as Chapter 18, "Soil Bioengineering for Upland Slope and Erosion Control," which was the first new chapter published in twenty years.

Locally available native woody plant cuttings are installed in the ground in specific configurations and served immediately to offer surface erosion control, soil reinforcements, horizontal drains, barriers to earth movement and hydraulic pumps or wicks. In the deeper installations, initially the cuttings function in much the same way as geogrids. As roots develop along the length of the embedded branches additional reinforcement and stability occurs. Live fascines are long bundles of branch cuttings bound together and installed in shallow trenches. They are able to immediately reduce surface erosion and riling. Overtime they root and produce top growth will both serve to increase their effectiveness. Woody vegetation, when properly designed and installed in these specific configurations, create stable surfaces and or composite earth masses.

Essentially the soil bioengineering method know as vegetated geogrid is a composite of engineering and vegetative components. The live fascines are used initially to mechanically stabilize the upper slope face and to provide drainage. This case study represents an excellent example of combined technologies.

CASE STUDY

Project Conditions

The project site is located on the Massachusetts Turnpike immediately adjacent to the eastbound lane at Mile 79.3 in Charlton, Massachusetts. The slope was approximately 145 meters (475 feet) in length, ranged from 3 meters (10 feet) to 15 meters (50 feet) in height and had a slope angle of approximately 1V:1.5H with vertical sections along the top caused by continual surficial erosion. These erosional failures ultimately formed a large exposed, unvegetated area that was increasingly vulnerable to a combination of progressive erosion and further failure caused by slope instability. Groundwater seepage, saturated surficial soils, and seasonal freeze-thaw cycles exacerbated the instability of this north-facing slope. While both sides of the area, were well vegetated and appeared stable it was apparent that the condition was expanding on both sides, as shown in Figure 1. A concrete barrier was also installed along the base of the slope to contain the eroded and failed soil preventing it from moving onto the roadway. Subsurface conditions at the site include widely graded, slightly cohesive, dense to very dense glacial till overlain by shallow surficial topsoil and forest mat. Bedrock was found to be within 3 meters (10 feet) of the slope base elevation.

Figure 1 View of the project from the west-bound lane, prior to slope excavation and reconstruction - September 1994

Historical Background

Due to the nature and extent of the failure conditions, the Massachusetts Turnpike Authority (MTA) decided it was imperative to correct this situation by reconstructing the slope. If not treated, these conditions would inevitably lead to further slope failures, additional maintenance costs and an expanding, unsightly, unvegetated slope along this scenic stretch of the Turnpike.

The aim of the project was to design and construct a 4V:1H earthen buttress immediately in front of the cut slope to provide internal, external, and compound stability. The soil bioengineering approach was adopted to meet the requirement of an aesthetically pleasing and environmentally sound reconstruction and to control surface erosion and internal surface drainage. This combined approach uses vegetated geogrids and live fascine to provide internal and surficial stability and to support long-term vegetative growth with little to no maintenance requirements. The geogrid is a hybrid design that incorporates live cut branches known as brushlayers (See Figure 2) in the frontal, wrapped portion of the reinforced soil slope. Over time, the live branches take root and bind the geogrid layers and soil together thereby increasing the internal stability of the slope.

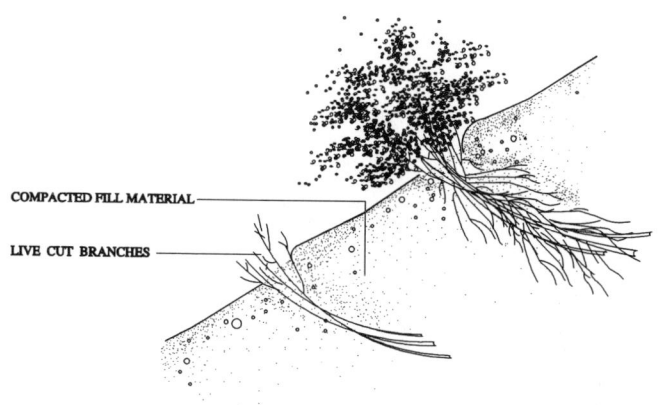

Figure 2 Illustration of a soil bioengineering brushlayer method

The combination of soil bioengineering and geogrid reinforcement provides the following benefits:

- Immediate slope stabilization and erosion control;
- No need to purchase additional rights-of-way as slopes can be reconstructed at very steep angles;
- Reduced maintenance costs because there is no need to return to the site to add soil or gravel, or to re-hydroseed, as the soils fill is all contained via frontal wraps;
- Modifies soil moisture regimes using back slope drainage systems. Additionally the brushlayer branches act as wick drains;
- Enhanced opportunities for wildlife habitat and ecological diversity; and
- Improved aesthetic qualities and scenic beauty through revegetation and naturalization of the slope.

Proposed Design

The design repair called for excavating back the failed slope approximately 6 meters (20 feet) to the same slope angle (4V:1H) as the proposed final slope and reconstructing a steepened, biotechnically stabilized earthen buttress. The slope was stabilized with layers of primary and secondary geogrids, burlap, cut vegetation, and live fascine on the top of the finished slope. Figure 3 shows a cross section depicting the failed slope (dashed line) and the remedial design.

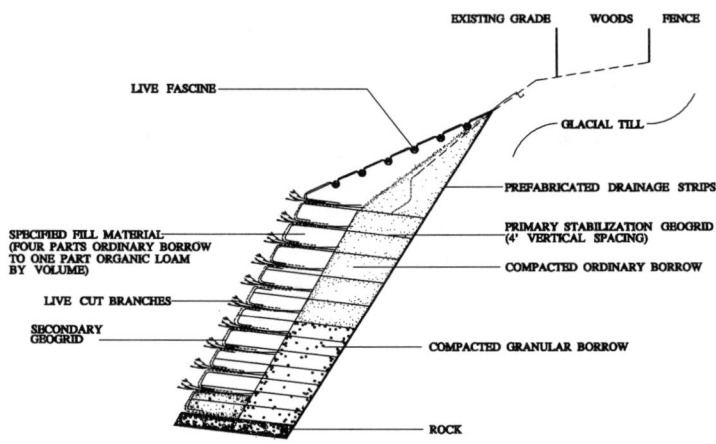

Figure 3 Cross section of the remedial slope design

The primary geogrid was designed to provide global, internal, and compound stability to the slope.

As shown in Figure 3, the secondary geogrid, and brushlayers (which together constitute the vegetated geogrid) were primarily designed to provide facial stability and erosion control to the slope face. As shown in Figure 4, the secondary geogrid is used to "wrap" the lift face between each row of brushlayer branches. The face wrap extends 1 to 1.75 meters (3 to 5 feet) into the slope. A front view of the vertical and lateral limits of the vegetated geogrid lifts is provided in Figure 5.

Figure 4 View of the vegetated geogrid on the front

Brushlayers consisting of 2.45 to 3 meter (8 to 10 feet) long willow (salix sp.) and dogwood (Cornus sp.) branches were placed on the constructed earthen terraces between each vertical lift. These brushlayers, which are installed in layers with the growing ends exposed, extend from 0.35 meters (1 foot) beyond the face approximately 3 meters (10 feet) back into the slope. During the growing season, these brushlayers will root and produce leaves, protecting the face of the slope from erosion. They will also provide some measure of internal stability initially and over the long term. The constructed earthen terraces and live branch brushlayers are shown in Figures 5 and 6.

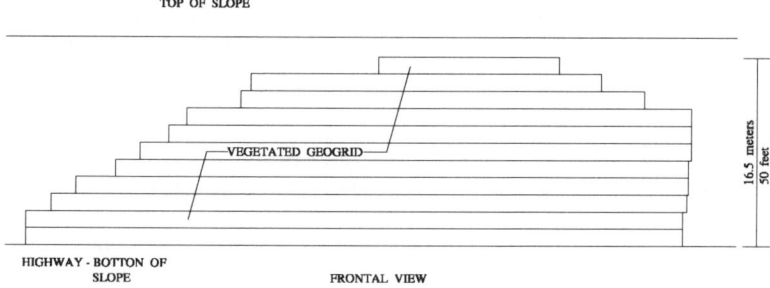

Figure 5 Frontal view of the as-built, illustrating the limits of the slope remediation

Figure 6 The alternating sequence of constructed earthen lifts terraces and live branch layer illustration

Drainage panels that extend vertically along the back side of the slope were designed to accommodate the migration of excess groundwater into the reinforced portion of the slope, preventing the buildup of hydrostatic pressures in that area. These panels connect into a crushed-stone drainage layer at the base of the slope, which extends the full length and width of the slope.

As shown in Figure 3, different types of backfill were used to reconstruct the slope in addition to the crushed-stone drainage layer at the base. Backfill constituted the structurally competent "core" while the specified fill at the face provides a media amenable to plant growth. The specified fill used in the front of each lift is a blended material consisting of four parts ordinary borrow to one part organic loam by volume. Nutrient soil tests were done on the imported fill materials, to establish their nutrient content levels. Fertilizers were needed to further optimize the growing conditions.

A 1V:3H fill was made above the steep slope and live fascine bundles (See Figure 7) were installed to offer immediate mechanical surface stabilization and drainage. Overtime the live fascines will rapidly revegetate the top of the slope with woody plant materials and offer additional reinforcement to the surface soil mantle. A live fascine is a collection of live cut branches grouped together in a bundle and secured with twine. The bundles were placed in trenches and anchored. In this particular application break the long 1V to 3H slope into a series of shorter slopes was most advantageous considering that it was a fill slope. Which connected to the natural land forms above and along the side. The bundles also act as drains, collecting water and transporting it laterally to both ends of the site. The surface soils in between the live fascine buddles were covered with a 100% natural woven coconut netting erosion control fabric known as coir with long straw mulch, grass and legume seed. The coir offers not only cover but some surface tension as well. It is installed under each live fascine bundle in the trench prior to live fascine installation. See Figure 7.

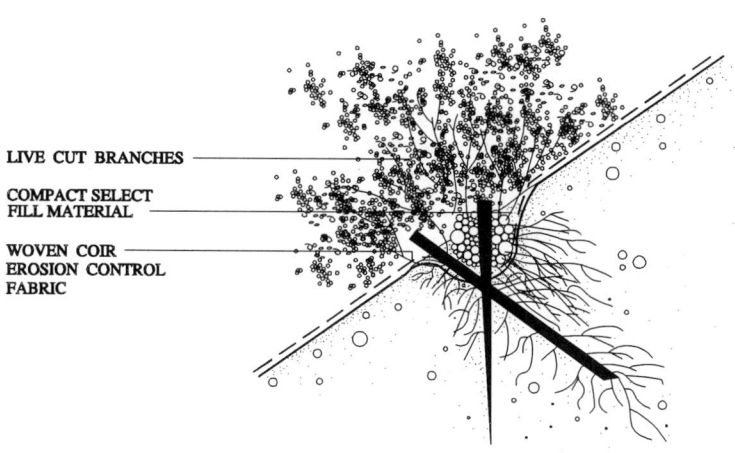

Figure 7 Illustration of an establishing live fascine soil bioengineering method with coir

Agronomic and Geotechnical Considerations

The design and construction of this slope presented several challenges in balancing the agronomic and geotechnical requirements. The factors to be balanced included: (1) the need to use organic matter in the slope to provide nourishment for plant growth and development versus the desire to construct the slope with free-draining, inorganic, granular soils alone; (2) the need for water and nutrients in the slope to sustain and promote vegetative growth versus the desire to remove water so as to eliminate hydrostatic pressures; and (3) the need to construct the slope during the fall and winter months with frozen soil. A final agronomic consideration was that the plant materials needed to be properly stored following harvesting to protect them from stress. A brief discussion of each of these topics follows.

- *Organic material* - To provide favorable conditions for plant growth, some organic material was needed in the backfill. To accommodate this, the outer 3 meters (10 feet) of slope, starting approximately 1.50 meters (5 feet) up from the base of the slope, was backfilled with a blended material consisting of four parts of ordinary borrow to one part loam by volume. By carefully selecting the fill mixture within the slope and checking the global stability and location of failure planes, it was possible to satisfy both the agronomic and geotechnical design requirements with one prepared backfill material.

- *Drainage* - To intercept some of the ground and divert it to the gravel subbase beneath the slope, prefabricated drainage strip systems were spaced on 4.50 meters (15 feet) on center the back cut glacial soils. This allows some groundwater to migrate into the reinforced zone without permitting hydrostatic pressures to develop. The willow and dogwood brushlayers also function as horizontal drains, reducing the possibility of hydrostatic pressures.

- *Construction time frame* - The freezing of subgrade soils that contain high organic material and water was a concern, since the construction was done mostly during the winter months. To minimize the impact of freezing, the lifts were adequately compacted at the end of each workday, and were inspected the following day for the occurrence of heave or formation of ice lenses. At times, accumulating snow had to be plowed from the previously constructed lifts before work could continue. It should be noted that a near record snowfall, in excess of 254 centimeters (100 inches), occurred during the winter this slope was constructed, following an unseasonably warm fall.

- *Harvesting, handling and storage of the branch cuttings* - The harvesting of suitable, biotechnically capable plant material and installation of needed soil bioengineering systems were carefully planned, coordinated and maintained. Harvest sites were located before the project started. Sites were then selected based on the quality of material, site accessibility, and proximity to the project site. These harvesting sites contained large stands of willow (Salix sp.), and Dogwood (Cornus Sp.) both species well-suited for soil bioengineering construction. Refrigerated trucks were used to transport and store the live cut branches on site, which allowed the cuttings to be stored for a month or more.(See Figure 8). Proper temperature and humidity controls were maintained to keep the branches in dormancy and prevent the cuttings from dying out. The use of refrigerated trailers allowed the contractor to transport larger quantities of material to the site, providing installation crews with immediate access to live cuttings when needed and improving overall operations and efficiency.

Figure 8 Photograph of a refrigeration van storing live branch materials

Figure 9 The condition of the project site, July 1996

CONCLUSIONS

The face of the major slope reconstruction, the vegetated geogrid, looked very good in the first season and has continued to develop. As shown in Figure 9, the slope is stable with no signs of upper slope or face erosion or failures. The vegetation is well developed. The fill slope above is also well vegetated and appears stable.

This soil bioengineering installations have restored function to this slope system and should continue to root and grow, thus providing for increased surface erosion control protection, soil reinforcement, and an aesthetically pleasing revegetated slope. Over time, natural invasion from the surrounding plant community is expected to occur, helping the system to further blend into the naturally wooded scenic setting of the area, satisfying the long-term mechanical and ecological goals of the project.

REFERENCES

Gray, D.H. and Sotir, R. (1995). Biotechnical and Soil Bioengineering Slope Stabilization. A Practical Guide for Erosion Control, John Wiley & Sons, New York, NY, USA.

Coppin, N., Barker, D., and Richards, I. (1990). Use of Vegetation in Civil Engineering. Butterworth's: Sevenoaks, Kent, England.

Sotir, R.B., Difini, J.T. and McKown, A.F. (1998) Soil Bioengineering\Biotechnical Stabilization of a Slope Failure, Presented at the Sixth Geosynthetics Conference on Geosynthetics, March 25-29, 1998, Atlanta, GA USA.

Performance Testing of Erosion Control BMPs at The ErosionLab

Dwight Cabalka, Member - ASCE[1]

Paul Clopper, Member - ASCE[2]

Abstract

In recent years, project owners and designers have shown increasing interest in obtaining reliable information on the physical and performance characteristics of erosion control materials. Users of these products consider test data as vital information to predict application success for a variety of projects, to evaluate potential materials and installation methods, and to establish project-specific quality requirements. Likewise, quality-conscientious manufacturers of erosion control products need this documentation for product research and development; sales and marketing purposes; installation guidelines; and product certification purposes.

Many users and manufacturers have done informal field tests on erosion control products with varying levels of documentation and reliability. Contract testing services are available through a few universities provided fee requirements and scheduling conflicts can be negotiated. As an alternative source, this paper details the design and construction of a dedicated, state-of-the-art erosion test laboratory located in Rice Lake, Wisconsin.

This outdoor facility includes twelve rainfall simulation plots for evaluation of erosion control materials on hillslope applications. Rainfall intensities of up to 254 mm/hr (10 in/hr) are achieved on the 2.4 m (8 ft) wide by 12 m (40 ft) long plots. In addition, the facility includes twelve open-channel flumes for evaluation of erosion control materials in open-channel applications. Velocities of up to 4.2 m/s (14 fps) and corresponding shear stresses of up to 480 N/m^2 (10 psf) are achieved at the maximum discharge of 1.7 m^3/s (60 cfs). Detailed information on the design criteria, selection of materials and construction methods is discussed. Unique features of the facility to document key external variables, including meteorological, geotechnical and agronomical conditions, as well as water quality aspects, are also discussed. Testing protocol and actual test results will be presented as this information becomes available.

Introduction

Documentation of product performance has recently become an issue of paramount importance in the erosion and sediment control industry. End-users and designers are more frequently requesting this information to provide the basis for construction project designs,

---
[1]National Applications Engineer, American Excelsior Company, 8601 F-5 West Cross Drive, #222, Littleton, CO 80123, 303-973-0417.
[2] Water Resources Engineer, Ayres Associates, 3665 JFK Parkway, #300, Fort Collins, CO 80525, 970-223-5556.

specifications and installations. Performance data is also valuable to manufacturers, so that they can provide realistic expectations to end-users, designers and installers. This information is central to product research and development; sales and marketing activities; installation guidelines; and product certification. With the development of a more sophisticated erosion/sediment control market, it is imperative that advancements in product performance testing be pursued.

Currently, information on performance is often difficult to secure or is of questionable accuracy. A few universities offer facilities for testing of erosion/sediment control materials and methods. However, comprehensive testing programs conducted at these facilities require a considerable amount of capital; often running into the tens of thousands of dollars. In addition to the high cost, university facilities are also typically scheduled many months in advance, making timely testing of erosion/sediment control products difficult. And there are no commercial organizations or test labs, accredited or otherwise, available at present to do performance testing of erosion/sediment control BMPs.

Informal field testing has therefore been the primary source of information on product performance. While actual field results amassed over a number of years may provide valuable insight on performance, it is often difficult to translate the results from one site to another. Job site conditions, such as soil, topography, weather and treatments, are often undocumented in field studies, yet can vary dramatically from site to site. The quality of materials and installation, also often undocumented, can have a significant effect, too, on the ultimate performance provided by products in actual field applications.

The ErosionLab, Rice Lake, WI

In order to provide timely and reliable information on the performance of erosion/sediment control BMPs, American Excelsior Company has recently sponsored the development of an in-house test facility near its manufacturing plant in Rice Lake, Wisconsin. The decision to construct this facility, known as "The ErosionLab", was based on three primary factors: 1) the need to evaluate the performance of conventional BMPs (blown straw, loose-placed rip-rap, silt fence, and hay bale checks); 2) the need for performance documentation on manufactured alternatives, including their installation methods, and; 3) the commercial drive to develop new solutions.

Although the Company has successfully furnished erosion control products to countless projects over the past thirty years, the need for carefully quantified data on performance in typical applications was determined to be a top priority. Since the mid-'80s, many competitive materials have entered the erosion control market; each with their own claims on performance - some realistic and some not. Aside from evaluation of the materials themselves, it was decided that related work, such as soil preparation, anchor patterns and termination details, should also be evaluated. There is currently very little data available on these related topics which in all likelihood have a significant effect on performance. And finally, the desire to improve existing BMPs and to innovate new solutions for erosion/sediment control applications was a major objective in establishing the ErosionLab.

The ErosionLab is tasked with determining the capabilities of a wide variety of products and methods to determine performance, as related to two basic criteria: 1) the ability to control erosion and reduce sediment loss prior to the establishment of vegetation, and; 2) the ability to accelerate seed germination and enhance the establishment of vegetation. Current testing is limited to evaluation of non-vegetated conditions, since: 1) all BMPs must initially perform in a non-vegetated condition to control erosion and retain seed; 2) the variables associated with vegetative establishment vary greatly depending on local climatic and agronomic conditions, and; 3) considerable time constraints are involved when attempting to evaluate vegetated conditions. By evaluating BMPs in unvegetated conditions, a conservative or "worst case scenario" is applied. Testing methodologies for evaluation of vegetated BMPs, under controlled conditions, are planned for the future. Generally-accepted scientific principles are applied to the test protocol, including documentation of existing conditions, use of control cases, certification of the materials/methods being applied, documentation of the test itself and collection of data.

The ErosionLab site comprises a portion of American Excelsior Company's 18.6 ha (46 ac) wood storage yard located near their manufacturing facility in northwestern Wisconsin. An on-site pond offers a ready source of clear water for conducting both rainfall and channel erosion studies. This 2 ha (5 ac) man-made pond was excavated on the site many years ago as the borrow source for an adjacent highway construction project. The pond has no surface outlet, receives inflows only from adjacent stormwater runoff (overland flow) and maintains its surface at the natural groundwater level. The on-site soil is predominantly Chetek sandy-loam, a fine- to medium-grained non-cohesive soil. Forested buffer strips separate the site from adjacent lands and the Red Cedar River.

To assure the integrity of this facility's design and subsequent operation, the Company selected an independent engineering consultant, Ayres Associates, to assist in the development of a Feasibility Study, including site selection, preliminary design, and budget analysis. This study was concluded by Ayres Associates' Fort Collins, Colorado office in early summer 1996 and accepted by the American Excelsior Company as the basis for development and construction of the laboratory.

As the result of this comprehensive study effort, The ErosionLab design incorporates the best elements of several existing research facilities, including the Jackhouse Reservoir Research Laboratory in England (CIRIA 1987), the Simons, Li & Associates overtopping flume facility in Colorado (Clopper & Chen 1988), the University of Hawaii (Sutherland 1996), Utah State University (Water Research Lab), the University of Minnesota (St. Anthony Falls Hydraulic Lab) and Texas A&M University (Texas Transportation Institute). Hillslope and channel applications are evaluated using The ErosionLab's Rainfall Erosion Facility (REF) and Channel Erosion Research Facility (CERF), respectively. In addition to these outdoor facilities, an indoor tilt-bed rainfall simulator [2.4 m W X 6 m L (8'W X 20'L)], materials testing laboratory and greenhouse facility are also included in this project, but are not detailed in this paper. Figure 1 is a site plan of The ErosionLab's outdoor facilities.

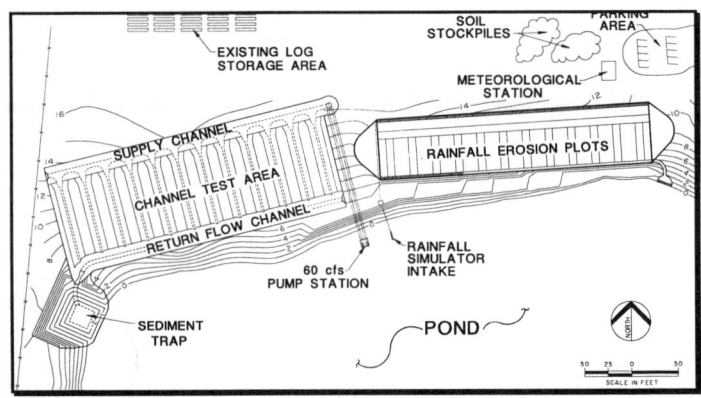

Figure 1. The ErosionLab site plan [CERF (left) and REF (right)].

Construction of the laboratory required approximately 1,800 m³ (2,400 cy) of cut and 1,725 m³ (2,300 cy) of fill, exclusive of the 45 cm (18 in) thick "veneer" material required for the test soils. Three veneer materials (loam, clay and sand) were selected, imported and installed, totaling approximately 2,100 m³ (2,800 cy). The REF development concept was based on simulation of rainfall to approximate the kinetic energy and erosive power of "typical" raindrops impacting a fairly steep hillslope. The CERF development concept used a water recycling theme where highly erosive flows are pumped from the pond through supply, test and return channels, and ultimately through a sedimentation basin with excelsior logs which filter the water before returning it to the pond.

The Rainfall Erosion Facility (REF)

To construct the REF embankment, excavated material from the channel area was placed in 15 cm (6 in) lifts and compacted using heavy earthmoving equipment to create a 3H:1V slope. The veneer soil treatments were placed on twelve separate plots, each 2.4 m (8 ft) wide by 12 m (40 ft) long, delineated with heavy-duty landscape edging to control run-on. The REF construction was completed with the installation of a 20 cm (8 in) diameter polyvinylchloride (PVC) pipe to convey water from a portable pump at the pond shore up to the top of the embankment. A manifold with 7.5 cm (3 in) diameter PVC risers was then constructed to distribute water to the top of each test plot.

To simulate natural events, eleven sprinkler risers, each 3 m (10 ft) tall with throw heights of approximately 1.2 m (4 ft), are positioned around the desired test plot to provide a 4.2 m (14 ft) fall height and uniform coverage. Valves on the sprinkler risers enable uniform intensities of 64, 128, 190, and 254 mm/hr (2.5, 5.0, 7.5, and 10.0 in/hr) to be delivered onto the test plots with a drop size typical of natural rainfall. At the maximum intensity, a discharge of about 3.8 l/s (60 gpm) is applied to the combined plot and overspray areas. Given a 20 minute test, a maximum REF "event" rains about 2,540 l (670 gal) of water on the plot. Figure 2 shows a typical rainfall simulator.

Figure 2. REF rainfall simulator.

Prior to testing, a thorough calibration routine is performed. This procedure documents the drop size distribution, fall height, rainfall areal uniformity and event intensity. Calibration and testing are not performed when the wind velocity is greater than 8 km/hr (5 mph).

To measure drop size distribution, three labeled pie pans are completely filled with sifted flour, struck off with a ruler to produce a smooth, uncompacted surface. Three supports approximately 20 cm (8 in) high (e.g., one-gallon cans) are located along the vertical centerline of the test plot, and at the horizontal quarter points. The filled pie pans are placed on the supports (horizontal, not parallel to the ground) and covered. At the desired test intensity, the cover is removed briefly so that drops impinge on the flour to form pellets. The pans are recovered after only a few seconds and before the drops start to touch each other. This procedure is repeated at each desired intensity. The flour pellets are air-dried for a minimum of 12 hours. Each sample of these semi-dry pellets is screened by emptying the entire contents of the pan onto a 70 mesh sieve to carefully remove as much loose flour as possible. The remaining pellets are then transferred to evaporating dishes and heated in an oven at approximately 43°C (110°F) for 2 hours. The total weight of the hard flour pellets is recorded. The pellets are sieved through standard soil sieves by shaking the stack for 2 minutes. Foreign matter and any double pellets are culled from each sieve, and the total weight and pellet count for each size are recorded.

Raindrop fall height is determined by measuring the average wetted height of the raindrop trajectory using a surveyor's rod. The rod is held vertically in the spray of a single riser and the wetted height measured. The height measurement is repeated for each desired intensity.

Twenty rain gauges are used to collect rainfall at each of the four target intensities to assure a uniformity of coverage and intensity. The average rainfall intensity over the entire test plot is the average depth of rainfall divided by the elapsed time of the test. The formula to calculate intensity (in centimeters per hour) is:

$$i = 60 \left[\sum_{j=1}^{J} P_j \div Jt \right]$$

Where:
i = rainfall intensity (cm / hr)
P_j = depth of rainfall (cm)
J = number of rain gauges (20 in this case)
t = time of test (minutes)

The Christenson Uniformity Coefficient is calculated for each intensity as:

$$C_u = 100 \left[1.00 - \Sigma |d| \div n\overline{X} \right]$$

Where:
C_u = Christiansen Uniformity Coefficient
$d = X_i - \overline{X}$
n = number of observations (20 in this case)
$\overline{X}$ = average depth caught
X_i = depth caught in each rain gauge , i

Based on the measured drop size, velocities can be determined for fall heights of approximately 4.2 m (14 ft) (Laws 1941) and compared with the terminal velocity of distilled water droplets in still air (adapted from Gunn and Kinzer 1949).

The kinetic energy imparted by the rainfall simulators at the soil surface can be determined by summing the kinetic energy of each drop size class multiplied by the relative percentage of that drop size, as determined by the distribution data. The kinetic energy represented by each size class is:

$$KE = 0.5mv^2$$

Where:
KE = Kinetic energy of drop size class
m = mass of drop
v = velocity of drop at the soil surface

As described in the USDA Agriculture Handbooks #537 and #703, the Erosion Index (EI) for the rainfall simulators is calculated as for the desired intensities:

$$EI = I \times 1099 \times [1 - 0.72 \exp(-1.27 \times I)]$$

Where:
EI = erosion index

I = rainfall intensity, in / hr

Once calibration is complete, measurements of runoff rates and sediment yields are made by collecting the sediment-laden runoff from each test plot at predetermined time intervals during the test. The collected quantities of sediment (i.e., soil loss) and runoff are then measured. Erosion patterns on the plot are also documented. The infiltration and runoff rates, and erodibility of the corresponding bare soil control are compared to the test BMP to determine its performance capability. Subjective observations are also made to document excessive material elongation, loss of material, etc.

Using the collected data, the following erosion analysis are performed:
- Runoff hydrographs, as discharge vs. time;
- Sediment concentration curves, as concentration vs. time;
- From the total runoff volume, the equivalent runoff Curve Number (CN), as used in the Soil Conservation Service (SCS) Curve Number method (SCS 1956) is calculated;
- From the peak runoff rates, the rational coefficient (C), as used for computation of peak discharge in the Rational Runoff Equation (Linsley, et al, 1972) is computed, and;
- From the total sediment yield, the equivalent Cover factor (C) used to compare soil loss to bare soil conditions in the USLE and RUSLE methodologies (USDA 1978 and 1997) is determined.

The REF design and testing protocol was carefully chosen for its simulation of natural rain events and accurate representation of many "real world" projects, including landfill caps, highway embankments and levees. A variety of BMPs are slated for investigation at the REF, including pnuematically-applied mulch (straw), hydraulically-applied cellulose mulches, Erosion Control Blankets (ECBs) and Turf Reinforcement Mats (TRMs).

The Channel Erosion Research Facility (CERF)

To construct the CERF, two vertical-lift pumps with a combined maximum discharge of 1.7 m^3/s (60 cfs or 27,500 gpm) were mounted on pre-cast concrete sumps. These pumps are driven by two 6-cylinder, 149 kW (200 hp) diesel engines. Water is pumped from the pond to the inlet control structure which feeds the supply channel where 23 pre-cast concrete headgates, each weighing 8,600 kg (19,000 lb), direct the tranquil flow into the desired test channel. In a similar manner to the REF, the 27.3 m (90 ft) long test channels are over-excavated to allow plating of the veneer soil treatments to a thickness of 45 cm (18 in). A trapezoidal cross-section is then created using a 0.6 m (2 ft) bottom width and 2H:1V side slopes, and 5 and 10 percent bed slopes (6 each). The CERF can achieve maximum velocities and shear stresses of about 4.2 m/s (14 fps) and 480 N/m^2 (10 psf). Figure 3 shows the test channel cross-section.

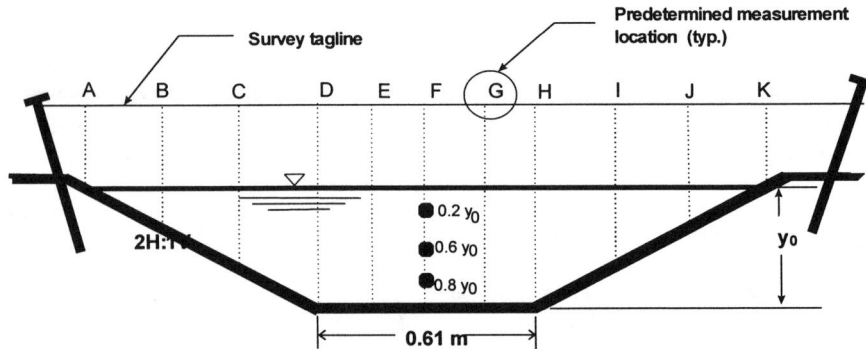

Figure 3. CERF test channel cross-section.

Centerline flow depth and velocity are measured at nine equally-spaced cross-sections located in the middle 12 m (40 ft) long reach of the test channel. Erosion patterns in the test channel are also documented. Pre-and post-test contour maps of the channel surface are developed using total station surveying equipment which downloads data into an earthwork software package and allows precise measurement of degradation or aggradation along the test reach. The primary performance indicator for channel tests is the Clopper Soil Loss Index (CSLI) which divides the total eroded volume by the total wetted surface area to determine an average erosion depth. Subjective observations are also made to document excessive material elongation, tearing, loss of material, etc.

Calibration of the CERF to determine the desired discharge is performed using the weir and the velocity-area equations. The weir equation measures the total head as the elevation difference between the measured water surface and the bottom of the weir. With supercritical flow in the steep test channel downstream from the weir, the total discharge is computed as:

$$Q = 1.65(L)(H)^{3/2}$$

Where: Q = discharge, m^3/s
L = width of weir, m
H = total head, m

Based on the trapezoidal cross-section in the water supply channel and the 3-point cross-section velocity measurements, the discharge in the channel using the velocity-area equation is computed as:

$$Q = V_1 A_1 + V_2 A_2 + V_3 A_3$$

Where: Q = discharge, m^3/s
V_n = measured velocity at each location, m/s

A_1, A_3 = flow area, m^2
Given:
Bottom width = 1.8 m, and side slopes = 2H:1V;
Flow areas, A_1 & A_3 = $0.45y_0 + y_0^2$

A_2 = flow area, m^2
Given:
Bottom width = 1.8 m;
Flow area, $A_2 = 0.9y_0$

Using the calibration data, discharge curves can be plotted, as follows:
- Discharge vs. engine speed (rpms)
- Discharge vs. hydraulic head in control structure
- Discharge vs. hydraulic head at each test channel

Once calibration is complete, analysis of the test data for the following variables can be determined: total discharge, velocity, flow depth, energy slope, resistance coefficient (Manning n-value) and boundary shear stress. Total discharge is determined by use of the calibration curves established for the given test channel and is also computed at each of the nine measurement cross-sections by the continuity equation, as follows:

$$Q = V_{avg} A$$

Where:
Q = discharge, m^3/s
V_{avg} = avg of the 3 centerline velocity measurements, m/s
A = cross-sectional area of flow, m^2

The energy slope, S_f, is determined by fitting a regression line through the energy grade line elevation determined at each of the nine measurement cross-sections, as follows:

$$EGL = WSE + V_{avg}^2/2g$$

Where:
EGL = energy grade line
WSE = water surface elevation
V_{avg} = average velocity, fps
g = gravitational constant, $62.4 f/s^2$

The Manning resistance coefficient, n, for each test is calculated, as follows:
$$n = R^{2/3} S_f^{1/2}/V_{avg}$$

Where:
n = Manning resistance coefficient
R = hydraulic radius, m*
S_f = energy slope
V_{avg} = average velocity, m/s

*Hydraulic radius equals flow area (m^2) divided by wetted perimeter (m).

The average and maximum boundary shear stresses are determined, as follows:

$$\tau_{avg} = \gamma \times R \times S_f \text{ and } \tau_0 = \gamma \times y_0 \times S_f$$

Where: τ_{avg} & τ_0 = avg & max shear stress, respectively, N/m^2
  γ = unit weight of water ($1001.6\ kg/m^3$)

The final step in analyzing channel erosion, the Clopper Soil Loss Index (CSLI), is calculated from the topographic data gathered before and after test flows using the total station equipment. The change in channel topography is used to define the performance of the lining material. Areas of degradation (soil loss) are quantified as "cut" and areas of aggradation (sediment deposition) quantified as "fill". Commercially-available earthworks software is used to evaluate the channel topographies and determine areas of cut and fill. The CSLI is calculated, as follows:

$$CSLI = C_T / A_T \times 100$$

Where: $CSLI$ = Clopper Soil Loss Index, cm
  C_T = total cut, m^3
  A_T = Wetted channel area, m^2

The CERF channel designs and testing protocol were also carefully selected to model natural stormwater events and many real world projects, including highway ditches, landfill drainageways and stormwater facilities. A variety of BMPs are slated for investigation at the CERF, including loose-placed rip-rap, ECBs and TRMs. Figure 4 is an aerial photograph of The ErosionLab.

Figure 4. Aerial photograph of The ErosionLab (10/97).

Data Analysis and Evaluation

Strict adherence to documentation and reporting procedures, including a thorough description of pre-test versus post-test conditions, is critical to the success of this program. This fundamental precept pertains to control runs, as well as tests of specific BMPs, and to channel tests, as well as hillslope rainfall simulations.

To improve the quality and assure the accuracy, Procedures Manuals for both the REF and CERF were developed and then reviewed informally by a number of independent, third-party agencies, including a number of universities and DOTs. Comments and suggestions received as a result of these reviews have been incorporated into the manuals. These protocols have been re-formatted and are currently being balloted in ASTM D-18.25.

To assure accurate and reliable assessment of performance, the testing methods, data collection procedures and analytical techniques for studies performed at The ErosionLab adhere to accepted scientific and statistical principles. Multiple independent tests are performed for control plots, as well as specific treatments, in order to assess the reproducibility of measured results. Comparisons are then made between the baseline condition and the BMP to quantify the degree of erosion/sediment control provided.

Summary

The ErosionLab represents a significant investment in the development of reliable technical information on product performance for typical applications; hillslope erosion due to rainfall and channel erosion due to concentrated flows. The data generated should be invaluable to development of performance-based standards within industry recognized organizations, including ASTM, IECA and AASHTO. But perhaps the most significant contribution of this state-of-the-art facility will be to offer the erosion control industry an ability to establish realistic expectations of performance that can be translated and applied to problem sites throughout the world. And by maintaining an end-user perspective, the solutions with the best combined value can be utilized.

References

Cabalka, D.A. and Clopper, P.E.
1997 "Design, Construction and Operation of a State-of-the-Art Erosion Technology Test Laboratory," Proceedings of the International Erosion Control Association, v. 28, pp. 341-350.

Cabalka, D.A., Clopper, P.E., and Johnson, A.G.
1998 "Research, Development and Implementation of Test Protocols for Channel Erosion Research Facilities (CERFs)," Proceedings of the International Erosion Control Association, v. 29, pp. 238-48.

Cabalka, D.A., Clopper, P.E., Johnson, A.G., and Vielleux, M.T.
1998 "Research, Development and Implementation of Test Protocols for Rainfall Erosion Facilities (REFs)," Proceedings of the International Erosion Control Association, v.

29, pp. 249-63.

Chen, Y.H., and Cotton, G.K.
1986 "Design of Roadside Channels with Flexible Linings," Hydraulic Engineering Circular No. 15, Report No. FHWA-IP-86-5, Washington, DC.

Clopper, P.E. and Chen, Y.H.
1988 "Minimizing Embankment Damage During Overtopping Flow," FHWA Report No. FHWA-RD-88-181, Washington, DC: Federal Highway Administration.

Construction Industry Research and Information Association (London, England)
1987 "Design of Reinforced Grass Linings." Construction Industry Research & Information Association (CIRIA).

Clopper, P.E. and Chen, Y.H.
1988 "Minimizing Embankment Damage During Overtopping Flow," FHWA Report No. FHWA-RD-88-181, Washington, DC: Federal Highway Administration.

Gunn, R. and Kinzer, G.D.
1949 "The Terminal Velocity of Fall for Water Droplets in Stagnant Air," Journal of Meteorology, v. 6, pp. 243-8.

Holman, J.P.
1984 "Experimental Methods for Engineers," McGraw-Hill Book Company, 4th Edition.

Laws, J.O.
1941 Easurements of the Fall-Velocity of Water-drops and Raindrops," Transactions of the American Geophysical Union, V. 22, pp. 709-21.

McWhorter, J.C., Carpenter, T.G., and Clark, R.N.
1968 "Erosion Control Criteria for Drainage Channels," Conducted for the Mississippi State Highway Department in cooperation with the US Federal Highway Administration, by the Agricultural and Biological Engineering Department, Mississippi State University, State College, MS.

Sutherland, R.A., and Ziegler, A.D.
1996 "Geotextile Effectiveness in Reducing Interrill Runoff and Sediment Flux," Proceedings of the International Erosion Control Association, v. 27, pp. 394-405.

Texas Department of Transportation
1993 "Procedures and Evaluation Criteria for Flexible Erosion Control Blankets, Flexible Channel Lining Materials, and Hydraulically Applied Projects," Texas Department of Transportation in cooperation with the Texas Transportation Institute - Environmental Management Program, Austin, TX.

Thibodeaux, K.G.
1985 "Performance of Flexible Channel Linings," US Federal Highway Administration

Report No. FHWA/RD-86/114, Washington, DC.

US Department of Agriculture
1978 "Predicting Rainfall Erosion Losses," Agriculture Handbook No. 537, Washington, DC.

US Department of Agriculture
1997 "Predicting Soil Erosion by Water: A Guide to Conservation Planning with the Revised Universal Soil Loss Equation," Agriculture Handbook No. 703, Washington, DC.

Geosynthetically Reinforced Vegetation:
Providing an effective, economical and aesthetically pleasing alternative to rock riprap

Roy J. Nelsen
Manager of Technical Services
North American Green, Inc.
14649 Highway 41 North
Evansville, IN 47711
(800) 772-2040
email: rjnelsen@nagreen.com

Abstract

Geosynthetically reinforced vegetation affords design engineers an effective, more economical and aesthetically pleasing alternative to rock riprap for lining channels with high shear stress flow events. A channel located in Monroe County, Iowa, along State Highway S-65 is a prime example of this technology in action. Soil erosion continuously occurred on this site, causing environmental impacts from sediment transport and down stream sedimentation. Unreinforced vegetation had previously been employed as the channel lining but had proven to be unsuccessful, resulting in severe erosion and channel scour. The channel along S-65 would be exposed to extended duration, high shear stress flow events and immediate erosive forces from continuous flow occurring in the channel bottom. This paper will present calculations and analysis used in the selection of a vegetative channel lining reinforced with a composite turf reinforcement mat. Furthermore, it will focus on design aspects and functionality of the geosynthetically reinforced vegetation versus rock riprap.

Introduction

As a design engineer, you are well aware that costs and performance are important characteristics that must be taken into consideration when determining suitable materials for use as a channel lining. Costs and performance are paramount in any erosion control design with the public's demand for getting the most for their money. In today's world, other attributes must also be accounted for, including aesthetic appeal.

In this case history we will examine a channel located in Monroe County, Iowa, along State Highway S-65, and the multitude of benefits associated with the use of geosynthetically reinforced vegetation (soft armor) as a channel liner. These benefits include but are not limited to improved performance, reduced costs, and over-all improved aesthetic appeal of the site. Furthermore, this case study will focus on the immediate and long-term design aspects and functionality of the geosynthetically reinforced vegetation channel liner.

Project Overview

The predominant design concern for this location was the conveyance of flow from numerous storm events resulting in long duration flows with high shear stress and velocities. Soil erosion continuously occurred on this site, causing environmental impacts from sediment and its migration to down stream locations. The channel was initially designed to provide a drainage system for runoff from the highway and 150 acres of surrounding agricultural land. Previously, unreinforced vegetation had been employed as the primary means of erosion control but had proven to be unsuccessful, resulting in severe erosion, channel scour and the removal of large amounts of sediment.

Since no easy and economical solution was readily apparent for solving the diversity of problems on this site, a team was developed consisting of an assorted group of professionals (i.e. engineer, erosion control specialist, and wetland plants specialist) to analyze all options. Furthermore, the engineer felt that this location would provide an excellent test site for a side by side comparison of TRM reinforced vegetation and rock riprap. For economical reasons and to better understand the afforded protection of reinforced vegetative channel liners, the design specified that the majority of the channel be lined with reinforced vegetation.

A small section of 30.5 cm (12 in) rock riprap was designed and installed at the end of the channel to serve as a control for comparison purposes. The rock riprap would also function as a potential check measure should any difficulties or erosion occur from the reinforced vegetation upstream.

Design Tools

A software program developed by North American Green entitled "Erosion Control Materials Design Software Version III" was used to assist with the design and selection of a channel lining for this site. The software is based on the permissible tractive force procedures contained in the Federal Highway Administration's Hydraulic Engineering Circular #15 (FHWA HEC 15, Chen and Cotton 1988) entitled "Design of Roadside Channels with Flexible Linings". To further improve the software's accuracy in the design of vegetated channel liners, data and methods from the United States Department of Agriculture's Agriculture Handbook 667 entitled "Stability Design of Grass-Lined Channels" (AG HBK 667, Temple et al, 1987) have been incorporated. The calculations and analysis afforded by this program assisted the design engineer in the examination of numerous channel liner types, including unreinforced and geosynthetically reinforced vegetation, and rock riprap.

The software's utilization of calculations and design procedures from the FHWA's HEC 15 and AG HBK 667 enables analysis of shear stress on the vegetation as well as the underlying soil to determine the overall stability of the channel lining. The maximum permissible design methodologies in both these references assist in selecting a channel liner with a permissible shear stress value greater than the tractive (shear) forces generated in the channel under the specified discharge.

FHWA HEC 15

Tractive forces are computed in the software program through the use of Manning's equation and the maximum Tractive Force formula furnished in the FHWA HEC 15.

Manning's equation for use with English units (FHWA HEC 15, 1988):

$$V = 1.49 R^{2/3} S_f^{1/2}/n$$

where V is the average flow velocity (feet per second) in the channel cross section; n is the Manning's roughness coefficient for the channel lining; 1.49 is a numerical constant used in English unit calculations for Manning's equation. A different constant is required for values computed in metric units (Chow, 1959); R is the hydraulic radius, in feet, which is equal to the cross sectional area divided by the wetted perimeter; and S is the friction slope of the channel in feet per foot.

Maximum Tractive Force Equation in English units (FHWA HEC #15):

$$Td = Y \times d \times s$$

where Td is the maximum shear stress on a lining material in pounds per square foot; Y is the unit weight of water in pounds per cubic foot; d is the maximum flow depth on lining in feet; and s is the channel bed slope or energy gradient in feet per foot. Through the use of permissible shear stress and these equations design engineers may determine if a flexible channel lining is considered stable for a specified discharge or flow event.

AG HBK 667

Determining the stability of both the plant and soil components is necessary in the design and selection of any channel liner utilizing vegetation as the primary means of scour protection. Both these components must be analyzed in determining the suitability and stability through the entire functional life of the lining. The long-term erosion control function of the TRM is supplemental in a reinforced vegetation channel lining. The TRM must assist in negating tractive forces penetrating the vegetative cover and must provide reinforcement for the vegetation.

Calculations, equations and data from AG HBK 667 further augment the accuracy of the software for selection of vegetation and reinforced vegetation channel liners. AG HBK 667 affords methods to refine and improve the definition of retardance classes of vegetation (i.e. height, density, and types) within the software. Furthermore, this reference provides the processes for calculation and determination of those shear forces penetrating the vegetative cover and acting on the soil surface.

The soil under the vegetative lining is deemed stable if the permissible shear value of the soil is greater than the shear force penetrating the vegetation. Determining the shear force penetrating the vegetation and acting upon the soil surface is accomplished through the following relationship:

$$Te = (Y \times d \times s)(1 - Cf)(Ns/N)^2$$

where Te is the shear stress exerted on soil beneath the vegetation by flow penetrating the grass stand in pounds per square foot; Yxdxs is the maximum shear stress on a lining material as outlined above; Cf is the cover factor of vegetation determined by the vegetation's type and stem density; Ns is the roughness coefficient of the underlying soil; and N is the roughness of the vegetation components.

Composite Turf Reinforcement Mat

The software's use of permissible shear stress calculations and analysis provided assistance in the analysis of the 100-year design discharge flow event through the developmental stages of the reinforced vegetation channel lining. The subsequent analysis established the specification of North American Green C350, a composite TRM (C-TRM) as the appropriate channel lining material for both performance and economical reasons over 30.5 cm (12 in) rock riprap.

C-TRMs are a relative new class of permanent rolled erosion control product. C-TRMs differ from 100% synthetic TRMs by incorporating an organic matrix material into a permanent three-dimensional net structure. The C-TRM specified for this project utilizes a coconut fiber matrix stitch bonded to an ultraviolet stabilized permanent 3 dimensional netting structure **(see Diagram #1)**. The combination of both organic and synthetic components in this matting will afford immediate and long-term benefits through the projected life of the channel.

Diagram #1

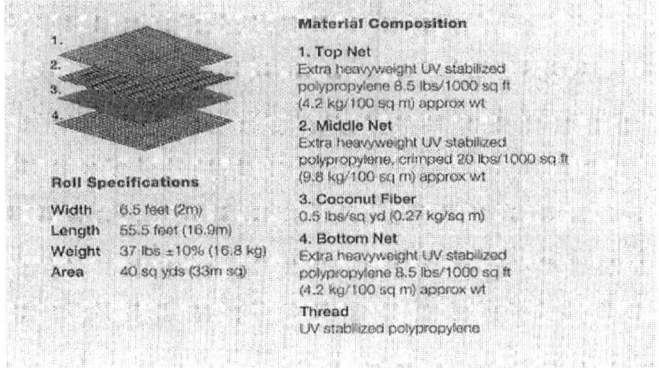

The coconut fibers have been selected for use in the C-TRM due to their diverse functionality. First, they afford scour protection for the channel's soil surface against flow induced tractive forces immediately after installation until the vegetation becomes fully established. Secondly, they function as an excellent mulching material to regulate environmental extremes at the seedbed, enhancing seed germination and vegetation establishment **(see figure #1)**. Finally, the coconut fibers provide a more economical alternative to 100% synthetic matrix materials.

Figure #1

The permanent three-dimensional synthetic net portion of the C-TRM affords a long-term system of reinforcement for the vegetation **(see Diagram #1)**. The netting structure of the matting increases the permissible shear stress of vegetated channel liners up to 384 Pascal (8 lbs/sqft, Urroz 1994, TTI, 1995), a level once thought only achievable with rock riprap or hard armor alternatives.

Problem:

Previous attempts at using unreinforced vegetation for the channel liner had failed at controlling the severe erosion in this channel **(see Table #1)**, but they did provide useful insight on three specific problems. The channel along S-65 would be exposed to extended duration (up to 12 hours) and high shear stress flow events occurring from discharges of 0.6 cubic meters per second (20 cubic feet per second (cfs)). Secondly, the erosion control measures must provide immediate protection with the presence of an almost continuous flow occurring in the bottom of the channel. Finally, diversion of drainage from across S-65, via a large culvert box, into this channel would result in large volumes of water entering the channel at a 45° angle. The ensuing turbulence and increased shear forces would further propagate scour action along portions of the channel lining.

Analysis of the channel design parameters and the unreinforced Retardance Class "C" vegetation revealed it would not provide a stable channel lining. HEC 15 defines a Retardance Class "C" vegetation as 15-30 cm (6-12 in) in height. Procedures from AG HBK 667 are then used in the software to further define the permissible shear stress for a Class "C" vegetation. The final permissible shear stress of a Class "C" vegetation is 202 Pascals (Pa) (4.2 lbs/sqft) **(see Table #1)**. Furthermore, the permissible shear of the underlying silt loam soil type (United States Department of Agriculture soil texture classification) of 1.68 Pa (0.035 lbs/sqft) is also defined in procedures from AG HBK 667.

Table # 1. Hydraulic Analysis for Unreinforced Vegetation

	Channel Bottom Width m (ft)	Left Side Slope (Horz. To 1)	Right Side Slope (Horz. to 1)	Channel Slope m/m (ft/ft)	
	2.7 (9)	3.0	3.0	0.100	
Discharge m³/s (cfs)	Peak Flow Period (hrs)	Velocity m/sec (ft/sec)	Area m² (sqft)	Hydraulic Radius m (ft)	Normal Depth m (ft)
0.6 (20)	12	1.20 (3.94)	0.47 (5.08)	0.13 (0.42)	0.15 (0.49)
Lining Type	Manning Coefficient	Permissible Shear Pa (lbs/sqft)	Calculated Shear Pa (lbs/sqft)	Safety Factor	Remark
Class C Unreinf. Veg.	0.067	Veg 202 (4.20)	145 (3.03)	1.38	Stable
		Silt Loam 1.68 (0.035)	2.0 (0.041)	0.85	Unstable

North American Green, 1995

The channel design parameters and flow conditions from the 100-year design discharge would result in calculated shear stresses slightly lower than the vegetation's permissible level. However, the calculated shear stress penetrating the vegetative cover was well in excess of the permissible value for the underlying silt loam soil type. The continued occurrence of erosion previously from this site further substantiates the rejection of unreinforced vegetation as the permanent channel lining.

Initially the engineer felt that 30.5 centimeter (cm) (12 inch (in)) rock riprap would be required to stabilize the channel against expected erosive forces from the 100-year design discharge **(see Table #2)**, but the installation costs of this option were prohibitive.

Table # 2. Hydraulic Analysis for 30.5 cm (12 in) Rock Riprap

Discharge m³/s (cfs)	Channel Bottom Width m (ft)	Left Side Slope (Horz. To 1)	Right Side Slope (Horz. To 1)	Channel Slope m/m (ft/ft)	
	2.7 (9)	3.0	3.0	0.100	
	Peak Flow Period (hrs)	Velocity m/sec (ft/sec)	Area m² (sqft)	Hydraulic Radius m (ft)	Normal Depth m (ft)
0.6 (20)	12	0.9 (3.02)	0.62 (6.63)	0.16 (0.51)	0.19 (0.61)
Lining Type	Manning Coefficient	Permissible Shear Pa (lbs/sqft)	Calculated Shear Pa (lbs/sqft)	Safety Factor	Remark
Riprap 30.5 cm (12 in)	0.100	192 (4.00)	183.4 (3.82)	1.05	Stable

North American Green, 1995

Three Phase Design System

It is noted that the extended velocity and shear stress resistance of reinforced turf is not realized until the grass lining and root systems are fully developed (generally two years) (CIRIA, 1988). The development of a reinforced vegetation channel lining occurs in three phases over approximately a two-year period, requiring the C-TRM to provide unique and varied forms of channel scour protection. The matting must vary its function from providing cover for a bare soil surface in Phase I to supplementing cover and reinforcing the vegetation in Phase III.

Specifically, Phase I requires the C-TRM to function as immediate cover reducing tractive forces from reaching the soil surface, while holding seed, soil and moisture in place to ensure germination **(see Figure #2)**.

Figure #2

During Phase II, both the channel and undeveloped plants are at a critical point, any damage to either during this period could potentially result in a channel failure. Thus, Phase II requires that the C-TRM maintain the same level and types of scour protection as in Phase I, while simultaneously protecting the developing root and stem systems of the vegetation, which are very susceptible to shear induced damage.

In the final phase (Phase III) of the reinforced vegetation channel lining the primary roll of the matting changes. With a mature vegetative stand in place and the coconut fibers fully degraded, the primary roll of the matting shifts from providing stand alone protection against channel scour to reinforcement of the vegetative stand. This function is accomplished through entanglement of the vegetation's root and stem structures during initial growth with the permanent three-dimensional net structure of the matting (i.e. root and stem reinforcement). The C-TRM must also partially function as a cover material to negate those shear forces penetrating the vegetation.

Solution:
Examination of the channel flow parameters under the design storm event determined that the C-TRM, with its coconut fibers present, would provide the necessary immediate (unvegetated) erosion protection during phase I of this channel **(see Table #3)**. The permissible shear stress of the C-TRM (108 Pa) equaled the calculated shear (108 Pa) generated by the discharge from the 100-year storm event.

Table # 3. Hydraulic Analysis for Unvegetated Composite TRM (Phase I)

Discharge m³/s (cfs)	Channel Bottom Width m (ft)	Left Side Slope (Horz. To 1)	Right Side Slope (Horz. to 1)	Channel Slope m/m (ft/ft)	
	2.7 (9)	3.0	3.0	0.100	
Discharge m³/s (cfs)	Peak Flow Period (hrs)	Velocity m/sec (ft/sec)	Area m² (sqft)	Hydraulic Radius m (ft)	Normal Depth m (ft)
0.6 (20)	12	1.68 (5.51)	0.34 (3.63)	0.10 (0.32)	0.11 (0.36)
Lining Type	Manning Coefficient	Permissible Shear Pa (lbs/sqft)	Calculated Shear Pa (lbs/sqft)	Safety Factor	Remark
Unveg. C-TRM* Phase I**	0.04	Veg 108 (2.25)	108 (2.25)	1.00	Stable

North American Green, 1995
* - Composite Turf Reinforcement Mat
**- Coconut fibers fully intact.

The longevity of the coconut fibers is accounted for in Phase II of the design software with an increased permissible shear on the underlying silt loam soil. The permissible shear for a C-TRM reinforced silt loam soil is 48 Pascal (1.0 lbs/sqft) in Phase III **(see Table #5)** and 108 Pascal (2.25 lbs/sqft) for Phase II **(see Table #4)**.

Table # 4. Hydraulic Calculations for Composite TRM Reinforced Vegetation (Phase II)

Discharge m³/s (cfs)	Channel Bottom Width m (ft)	Left Side Slope (Horz. to 1)	Right Side Slope (Horz. to 1)	Channel Slope m/m (ft/ft)	
	2.7 (9)	3.0	3.0	0.100	
Discharge m³/s (cfs)	Peak Flow Period (hrs)	Velocity m/sec (ft/sec)	Area m² (sqft)	Hydraulic Radius m (ft)	Normal Depth m (ft)
0.6 (20)	12	1.20 (3.94)	0.47 (5.08)	0.13 (0.42)	0.15 (0.49)
Lining Type	Manning Coefficient	Permissible Shear Pa (lbs/sqft)	Calculated Shear Pa (lbs/sqft)	Safety Factor	Remark
C-TRM* Reinf. Veg.	0.067	Veg 202 (4.20)	145 (3.03)	1.38	Stable
Phase 2**		Silt Loam 108 (2.25)	13 (0.27)	8.34	Stable

North American Green, 1995
* - Composite Turf Reinforcement Mat
** - Coconut fibers partially degraded.

Furthermore, the matting's permanent three-dimensional net structure would provide continued erosion control and the required reinforcement for the expected stand of vegetation **(see Table #5)**. The channel liner design was determined to be stable since the permissible shear stresses exerted on the reinforced vegetation (202 Pa) and soil (48 Pa) were well in excess of the calculated shear stress on these components (Reinf. Veg. 145 Pa and silt loam 13 Pa).

Table # 5. Hydraulic Analysis for Composite TRM Reinforced Vegetation (Phase III)

	Channel Bottom Width m (ft)	Left Side Slope (Horz. To 1)	Right Side Slope (Horz. to 1)	Channel Slope M/m (ft/ft)	
	2.7 (9)	3.0	3.0	0.100	
Discharge M³/s (cfs)	Peak Flow Period (hrs)	Velocity m/sec (ft/sec)	Area m² (sqft)	Hydraulic Radius m (ft)	Normal Depth m (ft)
0.6 (20)	12	1.20 (3.94)	0.47 (5.08)	0.13 (0.42)	0.15 (0.49)
Lining Type	Manning Coefficient	Permissible Shear Pa (lbs/sqft)	Calculated Shear Pa (lbs/sqft)	Safety Factor	Remark
C-TRM* Reinf. Veg.	0.067	Veg 202 (4.20)	145 (3.03)	1.38	Stable
Phase 3**		Silt Loam 48 (1.00)	13 (0.27)	3.71	Stable

North American Green, 1995
* - Composite Turf Reinforcement Mat
** - Coconut fibers fully degraded.

Performance

The performance of the C-TRM during the initial phase was critical because rainfall events resulted in an almost continuous discharge in the channel through installation of the matting and vegetation establishment. Flow depths of 15 cm (6 in) were not uncommon through the establishment phase of the reinforced vegetation channel lining **(see Figure #3)**. These flows tested the immediate erosion control capabilities of the unvegetated matting. The C-TRM worked well and did not allow erosion or loss of seed as was apparent by a dense stand of vegetation shortly after installation.

During Phase II, the site was impacted by a discharge that was estimated well in excess of the 100-year design storm event. The partially vegetated C-TRM with most of the coconut fibers still present protected the channel. There were no visible signs of erosion, rilling or undermining of the matting after subsidence of flow in the channel.

Figure #3

The coconut fibers also functioned as an excellent mulch material which improved seed germination rates and facilitated rapid vegetation establishment in the channel during Phase I and II. Budgeting constraints limited the C-TRM installation to the bottom and partially up the side slopes of the channel. In comparison, those areas of the channel protected by the matting and its coconut fibers displayed greater germination rates and more vegetative cover than those areas left unprotected **(see Figure #4)**.

Figure #4

Economics:

Providing a service that is both functional and economically sound is the basis for the fundamental theory of value engineering. Through the use of a C-TRM reinforced vegetation channel lining instead of rock riprap the design engineer for this site achieved increased performance at a lower cost. The economic savings incurred for this 6,700m^2 (8,000 sq yds) of channel lining were realized in reduced material costs and installation times. The average costs for 30.5 cm (12 in) rock riprap delivered and installed on this site was estimated at $9.00/m^2 ($7.50/sq yd). The selected C-TRM's installed cost was approximately $5.40/m^2 ($4.50/sq yd), resulting in a substantial total cost savings of $24,120 on this project.

Results:

Installation of the CTRM began immediately after channel construction activities. As mentioned previously the C-TRM was partially vegetated when a number of massive storms impacted the surrounding area. The subsequent runoff resulted in numerous high discharge flow events taking place in the channel prior to vegetation establishment. The storms appeared to be in excess of the 100-year design storm and resulted in high shear stress flow events. The reinforced immature vegetation provided the erosion protection for this channel even under these excessive flow events. The protection afforded by the C-TRM was apparent when the flows dissipated and there were no visible signs of soil loss or undermining of the matting **(see Figure #5)**.

Figure #5

The storm events had disastrous effects on the section of channel where 30.5 cm (12 in) rock riprap was installed **(see Figure #6)**. The high shear stress discharges caused the formation of scour holes, gully erosion and excessive soil loss in this area of the channel. This portion of the channel was actually eroded to the underlying bedrock in some areas. Most remarkably though was that a majority of the rock riprap was moved down stream from its initial location causing further economic impact **(see Figure #6)**.

Figure #6

Conclusion:

The Monroe County project is a prime example of how a C-TRM reinforced vegetative channel lining affords an economical and performance advantage over rock riprap. Through the use of a C-TRM on this project large economical savings were a direct result compared to the costs of a rock riprap channel lining. The geosynthetically reinforced vegetation channel lining provided performance levels once thought only achievable with hard armor alternatives. In fact, the C-TRM effectively protected the channel against the same erosive forces that dislodged and moved 30.5 cm (12 in) rock riprap from the channel.

REFERENCES

Chen, Y.H. and G.R. 1988. Cotton Design of Roadside Channels with Flexible Linings. U.S. Department of Transportation, Federal Highway Administration, Hydraulic Engineering Circular Number 15 (HEC 15).

Chow, V.L. 1959. Open-channel hydraulics. McGraw-Hill Book Co.

Hewlett, H. W. M., L. A. Boorman, and M. E. Bramley 1987. Design of reinforced grass waterways. Construction Industry Research and Information Association (CIRIA). Report #116.

North American Green 1995. Erosion Control Materials Design Software, Version III.

Temple, D. M., K. M. Robinson, R. M. Ahring, and A. G. Davis 1987. Stability Design of Grass-Lined Open Channels. U.S. Department of Agriculture, Agricultural Research Service, Agriculture Handbook Number 667 (AG HBK 667).

Urroz, Gilberto Ph.D., P.E. 1993 and 1994. Utah Water Research Laboratory, College of Engineering, Utah State University, Logan, Utah. Channel Liner Permanent Erosion Control/Turf Reinforcement Blankets. High Velocity Flow Tests of Two Root-Reinforcing Materials Under Bare and Sodded Conditions (1993). High Velocity Flow Tests of C-350 and P-300 Reinforced Sod (1994).

VEGETATION SELECTION FOR ROLLED EROSION CONTROL PRODUCT

Daniel Hunt[1], Deron N. Austin, P.E.[2] and William Agnew[3]

Abstract

Engineers pride themselves in their ability to solve complex problems with cost effective and technically sound approaches. When revegetating construction sites, the solution may be a temporary erosion control blanket on a slope or a turf reinforcement mat for long-term protection in a high flow channel. After selecting and installing such materials to inhibit erosion, one must realize that only a portion of the job is completed. The selection of appropriate vegetation to ensure permanent stability of the surface soils still needs to be achieved. Questions frequently arise such as: what criteria do I use to select vegetation and what vegetation do I select? Should I use plants that are indigenous to the area or introduce species that may be more suitable to the application? Are the soils suitable to maintain long term vegetation establishment? These questions are commonly asked by the engineering community and often baffle the most competent engineer.

This paper is designed to assist engineers in plant material selection and vegetation establishment for rolled erosion control products on construction sites. The paper will show properly selected, specified and installed vegetation is paramount to successfully complimenting other best management practices to ensure that the long-term solution to slope and channel stability is not an afterthought.

---

[1] Engineering Assistant, Synthetic Industries, Inc., 4019 Industry Drive, Chattanooga, TN 37416 USA.
[2] Manager of Engineering Services, Synthetic Industries, Inc., 4019 Industry Drive, Chattanooga, TN 37416 USA
[3] President, REVEG Environmental Consulting, Inc, a Division of Granite Seed Co, 1697 West 2100 North, Lehi, UT 84043 USA

Revegetation Components

In general, location, site geometry, contours and grading have the greatest impact on the revegetation plans. For permanent stabilization, however, more detailed information is required. The fertility and type of soil must be assessed. If unfertile, soil amendments & supplemental nutrients must be added prior to final seedbed preparation and tilling. Once prepared, an appropriate seed is selected and application rate is specified. Temporary degradable erosion control blankets (ECBs) are typically used to protect newly seeded areas on the site where raindrop impact and moderate sheet runoff is present. In areas where vegetation alone will not provide sufficient long-term protection against erosion, synthetic turf reinforcement mats (TRMs) may be specified. Like any permanent system, vegetation requires maintenance. These components, coupled with favorable environmental conditions, will lead to successful and attractive projects.

Site Contouring & Common Applications

Although we would like to think that construction activities disturb as little soil as possible, the reverse is usually true. As a result, contractors are required by the National Pollutant Discharge Elimination System (NPDES) regulations to submit an erosion and sediment control plan 14 days prior to mobilization for any site (five) 5 acres or greater. The plan is meant to outline the contractor's intentions to protect disturbed areas of the sites. These areas are typically identified as: (1) site perimeter; (2) slopes; (3) drainage swales and open flow channels; (4) bank protection; and (5) pipe inlet/outlet protection.

On an ideal site, the vegetation along the site perimeter is left undisturbed forming a grass buffer to capture sediment being transported in overland flow. Unfortunately, many sites are stripped of most vegetation and the perimeters must be physically protected and temporarily seeded. Soil preparation and seed selection is determined quickly by the contractor instead of natural resource professionals. Due to the function of grass buffers, the temporary seed should be selected, sown and irrigated so that vegetation establishes quickly and densely. In this application, slopes are gentle and overland flow is shallow.

Soil slopes have the potential to generate a tremendous amount of sediment-laden storm water. According to the Universal Soil Loss Equation, prompt final grading and permanent revegetation activities can reduce sediment by as much as 200 percent. This early investment by the contractor results in less on-site maintenance and attention from enforcement agencies. In general, south and west facing slopes are dryer and have soils that are less defined than slopes with either north or east facing aspects. If not horizontally reinforced with geotextiles, slopes should be graded to stable angle that permits mowing equipment to safely operate. In most instances, slopes are 3H:1V and flatter and upland runoff is diverted around or through the structure.

Storm water runoff is routed through excavated swales and channels for conveyance to detention basins. Since they are typically limited in size and effectiveness, reducing sediment load through the use of reinforced grass-lined systems will prolong the life and capacity of these basins. Research has shown the ability of grassed channels to enhance the deposition of sediment (Clary et. al. 1993). In fact, the presence of vegetation has been shown to increase sediment entrapment in some locations by as much as 700 percent (Abt et. al. 1996). Since they reinforce the root systems, TRMs can extend the performance limits of vegetation in swales and channels by more than 100 percent (Carroll et. al. 1991). Concentrated flows during peak storm events often generate velocities that exceed the limits of natural vegetation (1.5 m/sec).

In aquatic environments, soil banks are subject to rainfall impact, runoff and wave impact. These aggressive conditions may be complicated by rapid changes in water table elevation or inundation at the toe. If left unprotected, internal seepage can transport soil through the bank face. Erosion protection plans must therefore consider a wide variety of revegetation approaches including, but not limited to, aquatic grasses, woody plants, alkaline-tolerant and sand-stabilizing species. Live plantings may be required in order to achieve immediate protection. Not only are banks capable of generating a tremendous amount of sediment themselves but the proximity to water bodies mandate the use of sediment control devices such as turbidity barriers, plant rolls or coir mesh. While ideally graded to 3H:1V, steeper banks are not uncommon on shorelines of ponds, lagoons, and dams.

Protecting the backfill surrounding pipe inlet and outlet areas have traditionally been dominated by rock riprap and concrete. Since outfall and turbulence is avoided by most engineers, a simple collar consisting of a TRM and durable grass will provide an aesthetically pleasing system to resist scour and back splash. When properly installed, this system will protect the backfill and transition storm water into and out of the culvert. Entrance and exit velocities are similar to those found in open channels.

Temporary Grass Covers

As previously stated, vegetation establishment is the key component to the long-term success of stabilizing disturbed sites. Temporary erosion control plantings are selected for their ability to establish a quick cover to protect the soil surface from wind and water erosion. Temporary plant materials may include the use of grain crops such as wheat, annual rye, oats and others or a sterile hybrid that provides temporary cover but does not produce viable seed for future propagation. When selecting annual plant materials, it may be important to prevent plants from going to seed as they may become competitive with target perennial plant materials. The most common measures to eradicate annual plants is to mow the site prior to seed head formation. However, this technique may be prohibitive on steep slopes or in drainage channels. These sites favor the use of sterile annuals to provide initial vegetation cover and erosion protection, but do not produce viable seed and the potential for future competition with perennials.

Permanent Grass Covers

The selection of permanent vegetation should be based on a number of criteria. The plant species should be well adapted to the site (i.e. native) and be able to reproduce and sustain their populations for the design life of the project. It may even be necessary to evaluate adjacent undisturbed areas to assist in the development of an appropriate species list. When selecting permanent grass covers, the final composition of vegetation (grasses, forbs, shrubs and trees) must be considered. Ecological succession can be promoted by using a wide variety of plant materials that work to facilitate gradual changes in the plant communities over time.

Site-adapted or introduced perennial plants can also be effective in providing permanent vegetation cover. Introduced plants are often very competitive with native plants, frequently to the point of exclusion. The historical use of introduced plants such as smooth brome (*Bromus inermis*) or crested wheatgrass (*Agropyron cristatum*) is wide-spread and well-documented. However, the current trend is to recreate or mimic the development of native plant communities whenever possible. Additional resources to assist selection of plant materials for disturbed lands may include the following:

- Universities and their extension services;
- USDA Natural Resources Conservation Service offices;
- Published county soil surveys;
- USDA plant materials centers;
- Plant information networks established by various state and federal agencies;
- Commercial seed suppliers;
- Private consultants specializing in plant material selection and use; and
- Publication in professional journals and conference proceeding articles.

Woody Plants

The use of woody plants is often utilized to enhance the features of permanent grass covers on environmentally sensitive sites. Once established, shrubs and trees provide extensive soil holding capacity and shading to retard the forces causing erosion. They may also be necessary for providing cover for animals and winter forage that does not become buried by snow. Roots mechanically reinforce a soil by transfer of shear stress in the soil to the tensile resistance in the roots (Gray and Sotir 1996).

Although in wide spread use today, some shrubs and trees have proven difficult to re-establish. Many species propagate by sprouting and are difficult to start from seedlings while others need scarification or burning to initiate the germination cycle. Re-establishment of shrubs and trees is further complicated by potential depredation and competition from grasses and forbs (Draves and Berg 1978). Although newly emerging seedlings are very palatable and attractive to wildlife, they compete for water and nutrients with quicker establishing grasses and forbs. When transplanted, seedlings must survive the shock of a sudden change in environment. It takes approximately one to three years for newly planted seedlings to develop sustainable root systems capable of adequately satisfying the plant's needs. The soil properties will mandate the extent of shrub or tree diversity because of different rooting requirements between the species.

Climate

A number of climatic variables affect plant growth and should be evaluated. These variables include the following information, which is usually available from the National Climatic Data Center or the National Weather Service (Gray and Sotir 1996):

- Minimum, maximum and average air temperatures;
- Maximum ground surface temperature;
- Length of growing season;
- Rainfall total and seasonable distribution; and
- Drought duration and time of year.

Selection of plant materials for applications in different geographic regions is challenging. Plants that do well in Nebraska may not be appropriate for applications in Nevada or Pennsylvania. Further specific plant materials may have attributes that are applicable throughout differing geographic areas and may be characterized as having the following adaptations and tolerances:

- Drought tolerant;
- Sand stabilizing;
- Acid tolerant;
- Alkaline tolerant;
- Moisture/Flood tolerant and
- Salt tolerant

The map (Figure 1) and table (Table 1) below allow for vegetation selection based on 8 Climatic Zones. Once the correct zone has been identified on the map, then the seed selection can be performed by cross-referencing the climatic zone and the application from Table 1.

Figure 1. Zones for Grass Adaptation in the United States

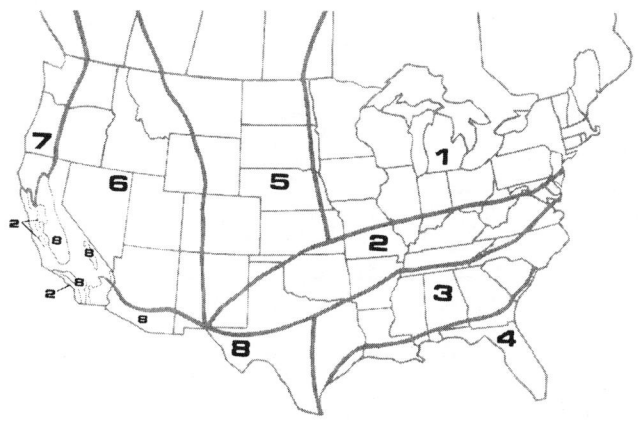

Table 1. Examples of Grasses

Zones of Adaptation	Species	Scientific Name	Typical Height (inches)	Retardance Class	Typical Applications			
					Site Perimeter	Slope	Channel	Bank	Pipe
Drought – Tolerant Bunchgrass										
5,6,8	Beardless Wheatgrass	Pseudoroegneria spicata ssp. Inermis	M	C – A	■	■		■	
5,6	Big Bluegrass	Poa ampla	T	A	■	■		■	
5,6	Bluebunch Wheatgrass	Pseudoroegneria spicata	M	C – A	■	■		■	
1,2,5,6,7	Hard Fescue	Festuca longifolia	M	C – A	■	■		■	
5,6,8	Indian Ricegrass	Oryzopsis hymenoides	M	C – A	■	■			
5,6,8	Needle and Thread	Stipa comata	T	A	■	■			
5,6,	Russian wildrye	Psathyrostachys juncea	M	C – A	■	■		■	
1,2,3,5,6	Sand dropseed	Sporobolus cryptandrus	M	C – A	■	■			
5,6	Siberian wheatgrass	Agropyron sibiricum	M	C – A	■	■		■	
1,2,5,6,8	Slendar wheatgrass	Elymus trachycaulus	M	C – A	■	■		■	
2,3,4,8	Weeping lovegrass	Eragrostis curvula	M	C – A				■	
1,2,5,6,7	Sheep fescue	Festuca ovina	M	C – A	■	■		■	
Drought – Tolerant Sod Forming Grass										
1,2,5,6,7	Canada Bluegrass	Poa Compressa	M	C – A	■	■		■	
1,2,3,5,6,7	Tall Fescue	Festuca arundinacea	T	A	■	■			
1,2,5,6,7	Creeping red fescue	Festuca rubra	M-T	C – A	■	■		■	
5,6	Pubescent wheatgrass	Elytrigia intermedia trichophorum	M-T	C – A	■	■		■	
5,6	Streambank wheatgrass	Elymus lanceotatus	S-M	E – C	■	■		■	

Zones of Adaptation	Species	Scientific Name	Typical Height (inches)	Retardance Class	Typical Applications				
					Site Perimeter	Slope	Channel	Bank	Pipe

Drought – Tolerant Sod Forming Grass, Con't

5,6	Intermediate wheatgrass	Elytrigia intermedia intermedia	M-T	C – A	■	■		■	
5,6	Thickspike wheatgrass	Elymus lanceolatus	M	C – A	■	■		■	
1,2,3,5,6,8	Western wheatgrass	Pascopyrum smithii	M	C – A	■	■		■	
3,4,8	Bermudagrass	Cynadon dactylon	M-T	C – A	■	■		■	
1,2,3,5,6,7	Timothy	Phleum pratense	M	C – A			■	■	■
1,2,5,6,7	Kentucky Bluegrass	Poa pratensis							

Sand Stabilizing Plants

5,6	Prairie sandreed	Calamovilfa longifolia	T	A	■	■			
1,2,3,4,5,6,8	Switchgrass	Panicum virgatum	T	A	■	■		■	
5,6,8	Sand bluestem	Andropogon hallii	T	A	■	■			
5,6	Indian ricegrass	Oryzopsis hymenoides	M	C – A	■	■			
5,6,8	Needle and Thread	Stipa comata	T	A	■	■			
5	Sand lovegrass	Eragrostis trichodes	T	A	■	■			
1	Beachgrass	Ammophilia spp.	T	A	■	■			
5,6	Blowout grass	Redfieldia flexuosa	M	C – A	■	■			
5,6,7	Sandhill muhly	Muhlenbergia pungens	M	C – A	■	■			

Acid Tolerant Grasses

1,2,5,6,7	Canada bluegrass	Poa compressa	M	C – A	■	■		■	
1,2,5,6,7	Perennial ryegrass	Lolium perenne	M-T	C – A			■	■	

GEOSYNTHETICS IN FOUNDATION REINFORCEMENT

Zones of Adaptation	Species	Scientific Name	Typical Height (inches)	Retardance Class	Typical Applications				
					Site Perimeter	Slope	Channel	Bank	Pipe

Acid Tolerant Grasses, Con't

Zones of Adaptation	Species	Scientific Name	Typical Height	Retardance Class	Site Perimeter	Slope	Channel	Bank	Pipe
1,2,5,6,7	Colonial bentgrass	Agrostis tenius	M	C – A			■	■	
1,2,5,6,7	Creeping bentgrass	Agrostis palustris	M-T	C – A			■	■	
1,5,6,7	Creeping foxtail	Alopecurus arundinaceus	M-T	C – A			■	■	
1,2,5,6,7	Hard fescue	Festuca longifolia	M	C – A	■	■		■	
3,4,8	Bermudagrass	Cynodon dactylon	M-T	C – A	■	■		■	
1,5,6,7	Meadow foxtail	Alopecurus pratensis	M-T	C – A			■	■	
1,2,5,6,7	Red fescue	Festuca rubra	M-T	C – A	■	■		■	
1,2,5,6,7	Redtop	Agrostis alba	T	A			■	■	
1,2,3,4,5,6,8	Switchgrass	Panicum virgatum	T	A	■	■			
2,3,4,8	Weeping lovegrass	Eragrostis curvula	M	C – A				■	

Alkaline Tolerant Grasses

Zones of Adaptation	Species	Scientific Name	Typical Height	Retardance Class	Site Perimeter	Slope	Channel	Bank	Pipe
5,6,7,8	Alkali sacaton	Sporobulus airoides	M	C – A	■	■		■	
5,6,7,8	Alkali grass	Puccinellia lemmoni	S-M	E – A	■	■		■	
2,3,4,8	Bermudagrass	Cynodon dactylon	M-T	C – A	■	■		■	
1,2,5,6,7	Perennial ryegrass	Lolium perenne	M-T	C – A	■	■		■	
5,6	Streambank wheatgrass	Agropyron ripadum	S-M	E – A	■	■		■	
1,2,5,6,7	Tall wheatgrass	Agropyron elongatum	T	A	■	■		■	
1,2,3,5,6,8	Western wheatgrass	Agropyron smithii	M	C – A	■	■		■	
5,6,8	Alkali cordgrass	Spartina gracilis	T	A					
5,6	Basin wildrye	Elymus cinereus	T	A	■		■	■	

Zones of Adaptation	Species	Scientific Name	Typical Height (inches)	Retardance Class	Typical Applications			
					Site Perimeter	Slope	Channel	Bank	Pipe

Alkaline Tolerant Grasses

Zones of Adaptation	Species	Scientific Name	Typical Height (inches)	Retardance Class	Site Perimeter	Slope	Channel	Bank	Pipe
5,6,8	Saltgrass	Distichlis stricta	S-M	E – A				■	
5,6	Russian wildrye	Elymus junceus	M	C – A	■	■		■	
5,6,7,8	Alkali grass	Puccinellia lemmoni	S-M	E – A				■	
1,2,5,6,8	Slender wheatgrass	Agropyron trachycaulum	M	C – A	■	■		■	

Grasses and Legumes Tolerant of Moist Soils

Zones of Adaptation	Species	Scientific Name	Typical Height (inches)	Retardance Class	Site Perimeter	Slope	Channel	Bank	Pipe
1,2,3,5,6,7,8	Alsike clover	Trifolium hybridum	M	C – A			■	■	■
5,6,8	Alkali cordgrass	Spartina gracilis	T	A			■	■	■
1,2,3,7	Reed canarygrass	Phalaris arundinacea	T	A			■	■	■
1,2,5,6,7	Colonial bentgrass	Agrostis tenuis	M	C – A			■	■	■
1,2,5,7,8	Creeping bentgrass	Agroslis palustris	M-T	C – A			■	■	■
1,2,3,7	Poa trivialis	Poa trivialis	M	C – A			■	■	■
1,5,6,7	Creeping foxtail	Alopecurius arundinaceus	M-T	C – A			■	■	■
1,5,6,7	Meadow foxtail	Alopecurius pratensis	M-T	C – A			■	■	■
1,2,5,6,7	Perennial ryegrass	Lolium perenne	M-T	C – A			■	■	■

Legumes

Zones of Adaptation	Species	Scientific Name	Typical Height (inches)	Retardance Class	Site Perimeter	Slope	Channel	Bank	Pipe
1,2,5,6,7	Crown vetch	Coromilla varia	M	C – A	■	■		■	
1,2,7	Birdsfoot treefoil	Lotus corniculatus	M	C – A	■	■		■	
2,3,8	Sericea lespedeza	Lespedeza cuneata	M	C – A	■	■		■	
1,2,3,5,6,7,8	White clover	Trifolium repens	S	E – C	■	■		■	
1,2,3,5,6,7	Alsike clover	Trifollum hybridum	M	C – A	■	■		■	

Height: Short (S) 1-12 inches; Medium (M) 13-24 inches; Tall (T) 25 inches or taller
Retardance Class: A is 24 inches or taller; B is 12 – 24 inches; C is 6 – 12 inches; D is 2 – 6 inches; E is 2 inches or shorter
Climatic Zones: 1 Cool – Humid; 2 Cool – Warm Season Transition; 3 Warm – Humid; 4 Tropical; 5 Cool Semi-Arid Plains; 6 Semi-Arid Inter-Mountain; 7 Cool – Humid; 8 Warm – Arid

Soil Properties

Soil conditions and properties vary dramatically throughout the United States as well as within a site. When not provided, soils information is available from the US Department of Agriculture's (USDA) Natural Resource Conservation Service (NRCS). Most NRCS mapping is characterized at an Order 3 level of reconnaissance with soil mapping scales of 1:12,000 to 1:250,000. When more precise delineation is required, Order 2 or Order 1 surveys are available. These soil surveys also contain vegetation data for various soil types.

Prior to and following disturbance activities, site-specific soil sampling and laboratory analyses should test a wide array of properties in order to make quality judgements concerning the plant growth medium. Field observations should include depth to hardpan or impervious layer, density, lithology, color, texture/structure, course fragments, presence of alkalinity or acidity and the native vegetation on the site. Laboratory analyses should determine grain size distribution, structure/texture, density, pH, exchangeable sodium, water repellency, moisture and nutrient content, trace elements and toxin levels. At a minimum, a field sample should be collected to evaluate potential soil amendment or fertilizer needs. Soil amendments may include the incorporation of lime in highly acidic soils, gypsum on highly alkaline soils, or organic matter to assist in providing good soil health and structure while tying up concentrated elements that may be harmful to plants. Soil conditions that may create challenges in re-establishing vegetation include the following (Gray and Sotir 1996; Agnew et. al. 1998):

- Soil materials on very steep, droughty, or unstable slopes;
- Shallow or stony soils;
- Strongly acid, strongly alkaline or high-salt soils;
- Soil material containing reactive pyrite;
- Soils with toxic materials such as excessive levels of arsenic, copper or aluminum;
- Soils very low in vital macronutrients (nitrogen, phosphorus or potassium);
- Continuously wet and ponded soils; and
- Absence of essential Rhizobium bacteria, mycorrhizal fungi or other beneficial microorganisms.

Mycorrhizal fungi and other beneficial microorganisms establish synergistic relationships with the fine roots of trees, shrubs, forbs and grasses. Through this relationship, mycorrhiza makes water absorption by the plant more efficient and improves absorption of essential mineral elements. Plants then exhibit significantly higher survival rates and lush growth, especially valuable in stressed environments.

Seedbed Preparation

Ideally and at a minimum, the physical condition of the soil surface must be developed to create a good seedbed to encourage vegetation germination and growth. A good seedbed would have the following qualities:

- Firm, but not compacted
- Relatively loose above the seeding depth to allow seed penetration and cover
- Free of weeds, large debris and large rocks
- Capable of holding moisture and at the same time be free draining

Favorable surface soil conditions are typically developed through cultivation by ripping and plowing compacted material, then disking and harrowing or raking the disturbance. If a surface is partially vegetated and not exhibiting the effects of accelerated erosion, it may be most effective to not cultivate and simply compliment the existing vegetation with other techniques to increase vegetative cover and production. Such techniques may include roughening the surface with dozer tracks, pitting, gouging and furrowing all of which slow runoff, minimize erosion, increase moisture retention and compliment erosion control material performance.

Planting Techniques

Seed is planted by a variety of techniques and include drilling, hand or mechanical broadcasting and hydraulic applications. By far, the most effective technique is drilling as this places the seed in the soil profile at depths advantageous to ensure germination and seedling vigor. It is important to drill the seed ¼ to ½ inch (5 - 10 mm) into the surface to prevent germinating seed from drying out. The seedbed bed should be firm to ensure adequate moisture holding capacity.

Broadcasting and hydraulic seed dispersion requires twice the amount of seed from an applications perspective and may not create an optimum growth medium for seedlings. Higher application rates may also be required when dealing with poor quality soils compared to favorable soil types. When broadcasting, the likelihood of wind, water or other natural factors transporting seed off site are dramatically increased. There lies an important role that effective erosion control applications provide, as they allow seed to be kept in place for a long enough period of time to ensure germination and encourage plant health. Broadcasting is the least expensive and fastest seeding method. When using hydraulically applied erosion control materials (mulches) in the semi-arid and arid regions of the western United States, it is imperative this be conducted in a two step application. First, the seed is dispersed on the soil surface by hand broadcasting or by using a tracer (light amount) mulch application to be followed with the full compliment of mulch material. This will ensure that the seed is in good contract with the soil surface and not suspended in the mulch material.

Seeding rates are based on the number of individual plants desired on any given area. Often this is determined by the number of pounds (kg) of pure live seed on a per acre (ha) basis of seeds (pure live seed or PLS) on a per square foot (m^2) basis. The formula for calculating the seeding rate is as follows:

$$\text{Planting Rate} = \frac{[A \cdot PLS]}{[q \cdot p \cdot g]}$$

A = (43,560 sqft or 1 acre (10,000sqm))
PLS = Pure Live Seed
q = number of seeds per pound (kg)
p = seed purity
g = seed germintation
Planting Rate = lbs of seed/acre (Kg/ha)

In general, the best season to plant is just prior to the season that receives the most reliable precipitation, usually spring or fall. The planting window can be expanded in the event that supplemental moisture is available to ensure plant growth needs. Typically, perennial grasses in cooler climates do best when planted in the fall of the year. Legumes and woody plants do best when planted in the spring and some annuals can be successful when planted in mid to late summer.

Rolled Erosion Control Products

A rolled erosion control product (RECP) is a temporary degradable or long-term non-degradable material manufactured or fabricated into rolls designed to reduce soil erosion and assist in the growth, establishment and protection of vegetation. Temporary degradable RECPs are composed of biologically, photochemically or otherwise degradable materials that temporarily reduces soil erosion and enhances the establishment of vegetation. A long term non-degradable RECP is composed of non-degradable materials that furnishes erosion protection and extends the erosion control limits of vegetation for the design life of a project. An erosion control net (ECN) is a planar woven natural fiber or extruded geosynthetic mesh used as a component in the manufacture of RECPs, or separately as a temporary degradable RECP to anchor loose fiber mulches. An open weave textile (OWT) is a temporary degradable RECP composed of processed natural or polymer yarns woven into a matrix, used to provide erosion control and facilitate vegetation establishment. An erosion control blanket (ECB) is a temporary degradable RECP composed of processed natural or polymer fibers mechanically, structurally or chemically bound together to form a continuos matrix. Finally, a turf reinforcement mat (TRM) is a long term non-degradable RECP composed of UV stabilized, non-degradable, synthetic fibers, nettings and/or filaments processed into three dimensional reinforcement matrices designed for permanent and critical hydraulic applications where design discharges exert velocities and shear stresses that exceed the limits of mature, natural vegetation. TRMs provide sufficient thickness, strength and void space to permit soil filling and/or retention and the development of vegetation within the matrix.

The installation of any RECP is very critical. The product must be installed in intimate contact with the prepared subgrade and anchored per manufacturers installation instructions. Once the vegetation begins to emerge, it must be maintained in order to maximize its appearance, effectiveness and health. This may include combinations of supplemental irrigation, periodic application of herbicides to control growth in right-of-way, mowing, etc.)

Conclusions

As discussed throughout this paper, engineers are frequently presented with and responsible for reclamation activities that are beyond their scope of expertise. Recently published EPA guidelines require more stringent sediment and erosion control for construction sites. Utilizing vegetation, as a Best Management Practice (BMP), in reclamation activities is one way to help meet these requirements and reduce sediment loads. The conditions for achieving successful reclamation are diverse and frequently require the input of a plant material professional in order to accomplish the necessary reclamation goals. Many services to assist in proper plant selection and application are available and my require only a phone call to positively effect the final outcome. Seeking out the expertise of natural resource professionals from the various state and federal agencies, the consulting community and from reliable and responsible erosion material manufacturer can make the difference between success or failure on the reclamation portion of your project.

References

Abt, S.R., Clary, W.P. and Thornton, C.I., "Ability of Streambed Vegetation to Entrap Fine Sediments", Proceedings of the American Institute of Hydrology, Interdisplanary Approaches to Hydrology and Hydrogeogeology, American Institute of Hydrology, Minneapolis, Minnesota, 1992.

Agnew, W., Theisen, M.S. and Schoal, H.R., "Disturbed Landscapes" in Time Savers Standards for Landscape Architecture. 2nd ed. McGraw Hill Publishing Co., 1998

Carroll, R.G., Rodencal, J. and Theisen, M.S., "Evaluation of Turf Reinforcement Mats and Erosion Control and Revegetation Mats Under High Velocity Flows", Proceedings of the XXII Annual Conference of the International Erosion Control Association, Orlando, Florida, 1991.

Clary, W.P., Abt, S.R. and Thornton, C.I., "Sediment Entrapment by Stream Channel Vegetation", Proceedings of the Management of Irrigation and Drainage Systems, Irrigation and Drainage Division, American Society of Civil Engineers, Park City, Utah, 1993.

Clary, W.P., Thornton, C.I. and Abt, S.R., "Riparian Stubble Height and Recovery of Degredaded Streambanks", Rangelands, Volume 18, Number 4, August, 1996.

Draves, R.W. and Berg, W.A., "Establishment of Mature Shrubs on Disturbed Lands in the Mountain Shrub Vegetation Type" Dept. of Agronomy, Colorado State University, 1978

Federal Register, Part II, Vol. 63, No. 6, published by Environmental Protection Agency, January 9, 1998

Gray, D.H. and Sotir, R.B, "Biotechnical and Soil Bioengineering Slope Stabilization – A Practical Guide for Erosion Control", John Wiley & Sons, Inc. New York, New York, 1996.

"Guide to Seed & Sod in the United States and Canada", published by Lofts Seed Inc, Winston-Salem, NC, 1991.

Subject Index

Page number refers to the first page of paper

Bearing capacity, 1, 49, 62
Biological operations, 92

Channels, waterways, 116
Construction methods, 103
Coverings, 77
Cyclic loads, 19, 34

Deformation, 34
Design criteria, 103

Earth reinforcement, 34
Erosion control, 77, 92, 103, 116, 130

Field tests, 77
Footings, 1, 49
Foundations, 1, 19
Full-scale tests, 34

Geogrids, 1, 19, 62, 92
Geosynthetics, 1, 34, 49, 62, 92, 116, 130

Laboratory tests, 103
Landfills, 77

Materials, 103

Performance evaluation, 103
Plasticity, 49
Preloading, 34
Prestressing, 34

Reinforcement, 1, 49, 62, 116

Sand, 19, 62
Selection, 130
Settlement analysis, 1, 19
Slope stabilization, 92
Soil stabilization, 130
Stress, 49
Stress distribution, 62

Transient loads, 19

Vegetation, 116, 130
Vegetative cover, 92

Author Index

Page number refers to the first page of paper

Agnew, William, 130
Akins, Ken, 1
Austin, Deron N., P.E., 130

Cabalka, Dwight, 103
Carson, David A., 77
Clopper, Paul, 103
Collin, James G., 62

Das, B. M., 19
Difini, John T., 92
Dodson, Robert, 62

Gabr, M. A., 62

Han, Jie, 1
Hunt, Daniel, 130

Koerner, George R., 77
Koga, Tetsushi, 34

McKown, Andrew F., 92

Nelsen, Roy J., 116

Sotir, Robbin B., 92

Tateyama, Masaru, 34
Tatsuoka, Fumio, 34

Uchimura, Taro, 34

Wayne, Mark H., 1

Zhao, Aigen, 49